HOW TO TURN ROUND A MANUFACTURING COMPANY

ELLIS HORWOOD SERIES IN
APPLIED SCIENCE AND INDUSTRIAL TECHNOLOGY

Series Editor: Dr D. H. SHARP, OBE, former General Secretary, Society of Chemical Industry; formerly General Secretary, Institution of Chemical Engineers; and former Technical Director, Confederation of British Industry.

This collection of books is designed to meet the needs of technologists already working in the fields to be covered, and those new to the industries concerned. The series comprises valuable works of reference for scientists and engineers in many fields, with special usefulness to technologists and entrepreneurs in developing countries.

Students of chemical engineering, industrial and applied chemistry, and related fields, will also find these books of great use, with their emphasis on the practical technology as well as theory. The authors are highly qualified chemical engineers and industrial chemists with extensive experience, who write with the authority gained from their years in industry.

Published and in active publication

PRACTICAL USES OF DIAMOND
A. BAKON, Research Centre of Geological Technique, Warsaw, and A. SZYMANSKI, Institute of Electronic Materials Technology, Warsaw
RECOVERY AND REHABILITATION OF CONTAMINATED LAND
R.J.F. BEWLEY, Biotreatment Limited, Cardiff, and D.H. SHARP, O.B.E., Sevenoaks, Kent
NATURAL GLASSES
V. BOUSKA *et al.*, Czechoslovak Society for Mineralogy & Geology, Czechoslovakia
INTRODUCTION TO PERFUMERY: Technology and Marketing
TONY CURTIS, Senior Lecturer in Business Policy and International Business, Plymouth Business School, and DAVID G. WILLIAMS, Independent Consultant, Perfumery and Director of Studies, Perfumery Education Centre
POTTERY SCIENCE: Materials, Processes and Products
A. DINSDALE, lately Director of Research, British Ceramic Research Association
MATCHMAKING: Science, Technology and Manufacture
C.A. FINCH, Managing Director, Pentafin Associates, Chemical, Technical and Media Consultants, Stoke Mandeville, and S. RAMACHANDRAN, Senior Consultant, United Nations Industrial Development Organisation for the Match Industry
THE HOSPITAL LABORATORY: Strategy Equipment, Management and Economics
T.B. HALES, Arrowe Park Hospital, Upton, Wirral
ASPECTS OF PRESSURE CONTROL AND SAFETY IN OFFSHORE DRILLING OPERATIONS
R. HOSIE, formerly Deputy Head, Mechanical and Offshore Engineering, Robert Gordon's Institute of Technology, now of Hosie Engineering Limited, Aberdeen
OFFSHORE PETROLEUM TECHNOLOGY AND DRILLING EQUIPMENT
R. HOSIE, formerly of Robert Gordon's Institute of Technology, Aberdeen
MEASURING COLOUR: Second Edition
R. W. G. HUNT, Visiting Professor, The City University, London
MODERN APPLIED ENERGY CONSERVATION
Editor: K. JACQUES, University of Stirling, Scotland
CHARACTERIZATION OF FOSSIL FUEL LIQUIDS
D.W. JONES, University of Bristol
BY-PRODUCTS AND WASTE MATERIALS IN FAT TECHNOLOGY
H. KIEWIADOMSKI, Technical University of Gdansk, Poland, and H. SZCZEPANSKA
PAINT AND SURFACE COATINGS: Theory and Practice
Editor: R. LAMBOURNE, Technical Manager, INDCOLLAG (Industrial Colloid Advisory Group), Department of Physical Chemistry, University of Bristol
ECONOMICS AND FINANCIAL STUDIES FOR ENGINEERS
D.J. LEECH, Senior Lecturer, Department of Management Science, University College of Swansea, Wales
CROP PROTECTION CHEMICALS
B.G. LEVER, International Research and Development Planning Manager, ICI Agrochemicals
HEALTH PROTECTION FROM CHEMICALS IN THE WORKPLACE
Editor: Dr P. LEWIS, Senior Executive, Occupational Health, Chemical Industries Association, London
HANDBOOK OF MATERIALS HANDLING
Translated by R.G.T. LINDKVIST, MTG, Translation Editor: R. ROBINSON, Editor, *Materials Handling News*. Technical Editor: G. LUNDESJO, Rolatruc Limited
FERTILIZER TECHNOLOGY
G.C. LOWRISON, Consultant, Bradford
NON-WOVEN BONDED FABRICS
Editor: J. LUNENSCHLOSS, Institute of Textile Technology of the Rhenish-Westphalian Technical University, Aachen, and W. ALBRECHT, Wuppertal
REPROCESSING OF TYRES AND RUBBER WASTES: Recycling from the Rubber Products Industry
V.M. MAKAROV, Head of General Chemical Engineering, Labour Protection, and Nature Conservation Department, Yaroslav Polytechnic Institute, USSR, and VALERIJ F. DROZDOVSKI, Head of the Rubber Reclaiming Laboratory, Research Institute of the Tyre Industry, Moscow, USSR

LANCHESTER LIBRARY

3 8001 00130 9701

35.00

LANCHESTER LIBRARY, Coventry University
Gosford Street, Coventry CV1 5DD Telephone 024 7688 7555

- 6 JUL 2005

- 7 DEC 2005

CANCELLED

- 5 JUL 2006

- 6 DEC 2006

- 5 DEC 2007

This book is due to be returned not later than the date and time stamped above. Fines are charged on overdue books

HOW TO TURN ROUND A MANUFACTURING COMPANY

BRIAN HALFORD WALLEY
Former Finance and Site Director,
Ferodo Ltd

ELLIS HORWOOD
NEW YORK LONDON TORONTO SYDNEY TOKYO SINGAPORE

Po 3829

First published in 1992 by
ELLIS HORWOOD LIMITED
Market Cross House, Cooper Street,
Chichester, West Sussex, PO19 1EB, England

A division of
Simon & Schuster International Group
A Paramount Communications Company

© Ellis Horwood Limited, 1992

Lanchester Library

All rights reserved. No part of this publication may be reproduced, stored in a retrieval system, or transmitted, in any form, or by any means, electronic, mechanical, photocopying, recording or otherwise, without the prior permission, in writing, of the publisher

Printed and bound in Great Britain
by Hartnolls, Bodmin

British Library Cataloguing in Publication Data

A Catalogue Record for this book is available from the British Library

ISBN 0-13-395922-8

Library of Congress Cataloging-in-Publication Data

Available from the publisher

Table of contents

Foreword	xi
Acknowledgements	xii
Preface	xiii

1 The manufacturing framework
1.1 Disaster and response—introduction   10
1.2 Company analysis   2
1.3 Changing course   6
1.4 The position paper   11
1.5 The production framework   11
1.6 Setting key objectives   18
1.7 Profit planning in a manufacturing company   19
    1.7.1 Target achievement   22
1.8 The role of management, especially the senior manager in the organization 23
1.9 Conclusion   24
Contention   25
A suitable set of strategies   26

2 The product market   27
2.1 Introduction   27
2.2 Pressures to change   28
    2.2.1 A business guru's change prediction   28
    2.2.2 An evaluative report on the DTI   28
    2.2.3 World automotive components product market developments 29
    2.2.4 The general response to changing market demands 30

		2.2.4.1 Strategic fit	30
		2.2.4.2 Obtaining strategic fit	32
		2.2.4.3 Strategic fit and the single market	35
		2.2.4.4 Niche marketing/Focused businesses/Focused factories	37
		2.2.4.5 Globalization	38
		2.2.4.6 The home market	39
	2.2.5	Market share	39
	2.2.6	Competitiveness	40
	2.2.7	High added value in product market strategy	41
	2.2.8	Competitor competence profiles	41
	2.2.9	The product range	42
	2.2.10	Forecasting	42
	2.2.11	Marketing/strategic fit activities	47
	2.2.12	Customer relationship—quality assurance	48
		2.2.12.1 Statistical process control	52
		2.2.12.2 Gaining quality awareness	53
		2.2.12.3 Quality circles	54
		2.2.12.4 The Deming philosophy and quality	55
		2.2.12.5 Quality—the Juran philosophy	56
		2.2.12.6 Taguchi	57
		2.2.12.7 Relationships with Japanese organizations	57
Contention			58
A suitable set of strategies			58
3 Technology			**60**
3.1	Introduction		60
3.2	High technology and the production process		62
3.3	Elements of CIM		65
	3.3.1	Why moving to CIM may be difficult	66
	3.3.2	Starting out on CIM	66
	3.3.3	Benefits of CIM	68
3.4	Advanced manufacturing technology in action		68
3.5	Material flow and technology		70
3.6	The Ferodo approach		70
3.7	Technology and the money problem		77
3.8	High technology—the only answer to manufacturing problems?		78
3.9	World-class manufacturers		79
Contention			80
A suitable set of strategies			80
4 Resource utilization and cost control			**82**
4.1	Introduction		82
4.2	People		83
4.3	Cash		85
	4.3.1	Cash forecasting	86

4.4	Reduction in administration costs		87
4.5	Privatization of fixed cost		89
4.6	Factory fixed cost		90
4.7	Working capital		91
	4.7.1	Stock/WIP	91
	4.7.2	Stock valuation	95
	4.7.3	Materials productivity	95
	4.7.4	Working capital generally	97
4.8	Waste disposal		98
4.9	Theft and security in resources control		99
4.10	Purchasing		101
4.11	JIT—a myth in its own lifetime		103
4.12	Research and Development		106
4.13	Product design and R&D		108
4.14	Resources data and control		109
Contention			110
A suitable set of strategies			111

5 Work organization and training   112

5.1	Introduction		112
5.2	Some organizational cul-de-sacs and responses		115
5.3	Goal conflict and work organization		117
5.4	Effective organization		119
5.5	Teams/groups		120
5.6	First-line supervision		122
	5.6.1	Introduction	122
	5.6.2	Objectives and performance measurement for first-line supervision	123
	5.6.3	First-line supervision in the 1990s	126
		5.6.3.1 Redundancy	126
		5.6.3.2 Recruitment	126
		5.6.3.3 Pay	126
		5.6.3.4 Discipline	126
		5.6.3.5 First-line supervision cohesion and team working	127
		5.6.3.6 Training and team working	128
	5.6.4	Conclusion	131
5.7	The road to setting up new work organizations		132
5.8	Trade unions		134
5.9	Discipline/unfair dismissal		136
5.10	Training		137
	5.10.1	General	137
	5.10.2	Why train?	138
	5.10.3	Communication	139
5.11	Senior and middle management training		141
5.12	Training for multi-skilling		142

Contention 146
A suitable set of strategies 147

6 Systems and information technology 149
6.1 Introduction 149
6.2 IT—definition and use 150
6.3 Systems design 152
6.4 Operational Planning 153
    6.4.1 Conflicts and organization 153
    6.4.2 Operational Planning—a uniting function 155
    6.4.3 Operational Planning—a suitable organization 155
    6.4.4 Principles/objectives 156
    6.4.5 Benefits of Operational Planning 156
    6.4.6 Optimized production technique/zone control 157
    6.4.7 Desirable improvements in Operational Planning 157
6.5 Systems specification/statement of user requirements 158
    6.5.1 What the users wanted from MRP II 159
    6.5.2 MRP II characteristics 159
    6.5.3 Systems specification outline format 161
    6.6 The initial debate 162
    6.6.1 Block diagrams 162
    6.6.2 Database 163
    6.6.3 Design of the core elements 167
    6.6.4 Other elements in MRP II 170
6.7 Implementation of MRP II 170
    6.7.1 The MRP II introduction · plan 170
    6.7.2 Some general points 172
    6.7.3 Conclusion 172
6.8 Management Accounting and business planning 173
    6.8.1 Introduction 173
    6.8.2 Line management involvement in Management Accounting development 174
    6.8.3 Elements of the system 178
    6.8.4 Top-down planning 185
6.9 Development of a Management Accounting system—a summary 185
6.10 Recording diagrams 187
6.11 Potential defects in Management Accounting 187
6.12 ABC—is this an answer? 191
Contention 191
A suitable set of strategies 192

7 Motivation and Reward systems 194
7.1 The dilemma in pay and motivation 194
7.2 Is money the only motivator? 197
7.3 Non-monetary motivation 197

7.4	Incentives and motivation—an approach	200
	7.4.1 Background	200
	7.4.2 Requirements	201
	7.4.3 Needs	202
	7.4.4 End result	202
7.5	Pay schemes—alternatives and problems	202
7.6	Management incentives/pay schemes	205
7.7	Relative deprivation and payment schemes	206
7.8	Objections to change and possible solutions	207
7.9	Management involvement in changing pay systems	209
7.10	The way forward	210
	7.10.1 Introduction	210
	7.10.2 Basis for change	210
7.11	Solutions	211
7.12	Future considerations	212
7.13	Conclusion	213
Contention		213
A suitable set of strategies		214

8 Formulating a manufacturing strategy for the nineties — 215

8.1	Overview	215
8.2	Culture in developing a manufacturing strategy	216
8.3	Company aims and objectives	220
8.4	Plan elements for a company in the nineties	220
8.5	What kind of a company must we be in the nineties?	227
8.6	Key questions	232
	8.6.1 General	232
	8.6.2 Product market	233
	8.6.3 Technology	234
	8.6.4 Resource utilization	235
	8.6.5 Work organization and training	235
	8.6.6 IT and systems	236
	8.6.7 Motivation and payment	236
8.7	State of the art	237
8.8	Key elements	238
8.9	Environmental analysis	239
8.10	Timescales	240
8.11	Profit planning	241
8.12	Conclusion	241

Appendix 1:	Ferodo	243
Appendix 2:	List of definitions and explanation of terms and acronyms used	244
Appendix 3:	Training	249

Foreword

This book describes the return of a major company from the brink of bankruptcy.

Its success depended upon the efforts of management at all levels, but a few simple lessons offer guidance for the management of industry in general.

Firstly, there were too many levels of management, resulting in second guessing and the demotivation of unit managers. Rectification of this, coupled with trust in local management and change where that trust was not justified, proved powerful stimuli.

Secondly, the computer had been used to provide excessive historical data with little useful forecasting. This led to a separation of top management from the problems of the business.

Thirdly, there was widespread overmanning which had to be addressed.

Fourthly, cash was not highly regarded or well controlled. It was essential to impose strict controls at the time and has proved to be a good motivator for management subsequently.

Simple monthly reporting from unit level was reduced to a single A4 sheet in which key financial indicators and regular forecasting of the annual out-turn were central. The effect of this on local management who were expected to perform and to control their units cannot be exaggerated.

Finally it should be noted that in spite of unsatisfactory profit and cash performance, we did not reduce the expenditure on research and product development; indeed we increased it. By this means, we accelerated the development of substitutes for asbestos and developed a wide range of projects.

The Company quickly returned to profitability and its share price rose eight-fold, permitting the acquisition of AE plc with a consequent widening of associated products and markets.

Central to these actions was a simplification of complex data and delegation of effective management to the operating unit.

Lord Tombs of Brailes
Chairman Rolls-Royce plc
Former Chairman of T&N plc

Acknowledgements

The author is grateful for permission to quote from publications made by the following organizations, journals and newspapers:

British Journal of Industrial Medicine
Department of Trade and Industry
The Economist
Engineering Employers' Federaton
Financial Times
Findlay Publications
Ford Motor Company Ltd Corporate Mission Statement
Hoare Govett Investment Research Limited. (The comments on Ferodo are part of a major review of T&N plc)
IBM
The Independent
Ingersoll Engineers
I.T. Europa Management
Jaguar
Lucas
Management Today
P A Consulting Group
Professional Engineering

Preface

Attempts to explain why organizations fade, fail and perhaps disappear for ever have given rise to many myths, created not least by those who have actually witnessed the process at first hand.

Senior managers blame the trade unions—they were greedy, would not cooperate, would not give up restrictive practices. The unions blame senior management—they were inept, never communicated, never invested. Other seemingly valid reasons are often quoted—foreign, especially Far Eastern, competition, lack of government support, short-termism by banks and the City.

Yet my experience points to only two major causes of disaster:

(1) the lack of will and perhaps ability of senior managers;
(2) the failure to address company culture and, especially, change in an appropriate way.

In *Production Management Handbook*[†] I proposed that companies should be regarded as systems within a framework of organizational elements. One element in the framework should not be changed without analysing the impact of the change on all the other elements.

In producing a manufacturing strategy *all* the elements in the framework, that is:

- The product market
- Technology
- Resources
- Work organization and training
- Information technology and systems
- Motivation and Reward systems

need to be reviewed as one entity.

In the last few years, there has been a barrage of advice—from the DTI, from consultants of all kinds, from academics, from magazines and books—for production managers to take up. Computer Integrated Manufacture, Manufacturing Resource Planning II, Computer-Aided Design/Computer-Aided Manufacture, Flexible Manufacturing Systems and a host of other potentially expensive words and phrases have come into current use. My experience, however, puts me firmly on the side of the NISSAN manager who said to stop worrying about progress through technology and start thinking about changes and improvements through people. It will cost less and get you further.

Of course new investment, possibly on a large scale, is essential; but technology alone should play a part, but only a part, in the making of the manufacturing phoenix. There is a chance that concentrating on Flexible Manufacturing Systems or MRP II or CAD/CAM, etc. will only reproduce the classic disasters of concentrating on one element in the production framework, which in the past has brought British industry major problems.

The eighties saw something of a resurgence in British manufacturing. Those companies that survived the recession of 1980–82 appear to have improved their efficiency. Managements scheme better; investment has improved. Even so, the downturn in the economy in 1990–91 has brought another crop of casualties. Companies have relearned the bitter strategy of borrowing heavily to fund major investments without considering all other relevant aspects of the production framework.

This book is unique. It derives from the personal experience of the author as a director of Ferodo Ltd, and from his work in helping to produce highly satisfactory results in a company which in the early eighties was in a very poor financial position. It is rare also in that there have been very few case studies written about British companies by the participants; this book is one of the few.

The book describes what Ferodo did in the eighties and, more importantly, what it still needs to do in the nineties if it is to 'take on the world'. The views and opinions, contentions and proposed strategies described in the book are, however, entirely those of the author. Such views and opinions do not necessarily reflect or coincide with those of Ferodo Ltd or T&N plc.

The author is also very aware that he was only one member of a team which introduced major changes in the company. It was a team effort which improved performance. For example, the author was not directly responsible for the significant engineering changes. These were conceived by Mr G. H. Briscoe (Works Director) and his team of engineers (led by Mr D. H. Vine) and much of the section on Technology reflects their ideas. (The opinions are my own.)

Members of Ferodo in the eighties—directors, managers and staff alike—should see this book as a record of their joint achievement.

---

† Published by Gower Press. Second edition 1985.

1

The manufacturing framework

1.1 DISASTER AND RESPONSE—INTRODUCTION

By the autumn of 1982, Ferodo Ltd and its parent company T & N plc were on the brink of disaster. There seemed a strong possibility that one or both would soon go out of business. However, the advent of Lord Tombs as the new Chairman, and the application of draconian measures to reduce costs and conserve cash, brought about a significant change.

As Ferodo set about initiating the toughest possible controls and the most radical re-structuring, there were many who wondered how we had ever got into such a deplorable position.

The decline of T & N and Ferodo was only symptomatic of a wider malaise in British industry. Indeed, manufacturing organizations throughout Western Europe and the USA were then under increasingly effective attack by low cost producers in the Far East or the Third World generally. What seem to be the key factors which have made the assault so successful? Some of the main ones appear to be as follows.

(1) The attackers had at least started out with lower unit labour costs. At the end of the eighties, South Korean workers were being paid between one-third and a half of the average pay for shop-floor workers in the UK. It has proved practically impossible for British manufacturing management to pay their labour force much less than the going rate in the local town hall or bank.

The (then) West Germans were even more out of line, with pay costs some 40% higher than those in the UK. (What they have since proved in part is that it is not labour rates that matter, but unit costs. If higher pay is given it must result in higher productivity.)

(2) The Japanese in particular had shown a first-rate ability to target markets, industries, even companies, where they could outmake and outsell products made

locally. A combination of good design, quality, service and price had enabled them to fight and win shares of the local market which provided suitable profit rewards. The short-term view had been eschewed for a well-considered appraisal of the market position required in five or more years' time.

(3) Good use had been made of reasonably advanced technology and systems and that more elusive characteristic 'management'. Japanese factories competing with Ferodo did not have significantly better production equipment. They just used it more effectively.

(4) As part of the implementation mentioned in section (3) there had been a studied build-up of procedures and skills such as total quality control, just-in-time stock control and quality circles. These by themselves might not have produced significant profit improvement or ensured survival in the longer term. They did, however, if appropriately applied, generate a different way of looking at and running a business. Total quality control, for example, motivated savings in raw materials and reduced the cost of re-work, but its more important impact could have been in the attention to detail and involvement in the production process by everyone, especially production operatives.

(5) The organization of the industrial units of these competitors was largely based on team work. Trade Unions, if they existed at all, had been friendly and cooperative with management. Everyone in the organization had been brought to understand and accept the same major goal the success of the enterprise. Restrictive practices had not been introduced, nor had strikes been fomented for trivial reasons.

(6) Working conditions and social benefits were often poor by the standards demanded by legislation in the West. The COSHH regulations would have received short shrift in many countries on the Pacific Rim. A production manager in the UK had to spend a considerable part of his time ensuring that he did not fall foul of either employment or health and safety law. This must have been a distraction from the main purpose of producing good quality, low cost products, on time.

(7) Our national culture had not been favourable to manufacturing and exporting of products. Survey after survey carried out in the UK showed that the nation at large was prepared to consume at a high rate, particularly of imports, but did not care overmuch whether British manufacturing industry lived or died. Individuals, especially the brightest graduates, did not consider working in British industry would give a satisfactory career.

No wonder the wealth creators in industry became disillusioned and cynical, noting in passing the crumbling schools and hospitals, inadequate transport facilities, and collapsing sewers, which their failure to create enough wealth had, in part, engendered.

1.2 COMPANY ANALYSIS

Many of these above-mentioned factors contributed in some part to the potential disaster we faced in 1982. However, no one should believe that a company by itself can have a major influence on such circumstances. Local management must concentrate on what went wrong in the company and what can be done to put it right. Certainly, past performance is a useful starting point. An analysis of environments is probably crucial. There could be, however, other factors purely related to one's company—product markets, technologies, the need for working capital, how good the planning and control is.

Area	What's wrong	What's right				
	Item	Data				Possibilities and comment
		Year 1	Year 2	Year 3	Year 4	
General economic	R.P.I. Average industrial pay Electricity Gas Rates					No company can exist without reference to the outside world factors which impinge on it. Companies should relate their performance to their environment
P & L items	Sales Variable cost Contribution Factory overhead Gross margins % Sales distribution/ Administration Exceptional expense Trading profit Interest Profit before tax					This data will provide some of the basis for the ratio analysis shown later. By itself the data may be ambiguous if not related to other factors

Balance sheet/
Capital employed
	Fixed assets:
	Land and buildings
	  Plant
	Working capital:
	  Stocks
	  Debtors
	  Creditors
	  Provisions					

Supporting ratios	Labour/Sales Material/Sales Sales/Factory /Fixed cost Sales/Admin.					These indicate trends in the use of labour and materials. Labour can be recorded as variable, fixed or SDA

4   The manufacturing framework   [Ch. 1

Area	Item	Data				Possibilities and comment
		Year 1	Year 2	Year 3	Year 4	
	Sales/Distribution Sales/Sales costs Payroll costs/ Employee					It would be useful to use Direct/Indirect/ Admin. etc. categories
	Days stock Raw Material WIP FGS					
Sales volume	Volume in pieces Prices					
General productivity	Labour cost Added value Material yield Energy/Sales					A good indication of materials productivity
Cash	Trading profit Provisions Depreciation Changes in working capital Interest Cash surplus					Cash generation is a vital factor in ensuring the survival of the company. The various cash-generating and -consuming items should be recorded and, where necessary, suitable action taken
Key statistics	Return on capital Operating profit/ turnover Turnover Capital employed Gross stocks/ Cost of sales Debtors/Sales Net working capital /Sales Payroll costs /Added value No. of employees Average profit per employee SDA expenses/Sales Cost of Sales/Sales					Key ratio in determining company efficiency Otherwise NET margin Measures effective use of stocks Key factor in generating cash A good measure of productivity, relating employee numbers to achievement Profit/Sales/Added value is useful in helping to determine productivity

Fig. 1.1. Company analysis.

We used the fairly simple analytical review shown in Fig. 1.1. The heading 'What's wrong, What's right' has to be seen in context. The 'context' in this instance is the history of the company, the strength of competition, the products made and markets served, the financial structure of the organization, plus, of course, financial requirements.

There is no easy answer to the 'What is our efficiency?' type of question. All that matters in the longer run is that shareholders are satisfied, that worldwide competition is challenged effectively, that the company is able to pay salaries which satisfy the management and workforce, and that no environmental pollution is caused.

Whether a company grows or not, what products it makes, which technologies it uses, even what view the City takes of it—all these things pale into insignificance so long as shareholders are reasonably happy and competition is at least matched and preferably beaten.

Normally, most production managers will see the monthly profit and loss statement and balance sheet regularly and also be bombarded by various management accounting report documents. Regrettably, these may all hide more than they reveal. Accountants are usually quite adept at making and losing profit when doing either is least expected.

The analysis we thought important was:

(a) Comparisons with the Retail Price Index (RPI).

Price rises. Ferodo, like many other manufacturing companies, has found it painful not to be able to raise product prices in line with UK inflation. No customer in, say, Germany, is likely to give a price rise which compensates for UK cost increases. Some products and product markets might have done better or worse than others: they need to be known, so that the best possible return is achieved.

Pay rises. Even in recent years, with an increasing emphasis on company performance, in setting pay and pay increases the demand of 'inflation plus $x\%$' has been the normal Trade Union request.

Variations in cost/RPI relationships. Some functions, activities and bought-in services may have a poor relationship with the RPI when compared with others. Knowing these is important.

Performance against Sales/Added value/Profit. Cost increases may be sustainable, if not totally acceptable, if their relationship with sales/added value/ or, say, profit improves. A 10% rise in direct pay which achieves a 15% increase in added value is obviously acceptable. The converse situation of, say, a 12% increase in pay and only a 5% improvement in contribution needs to be considered very carefully. Costs should always be related to the changes/improvements they should achieve.

(b) Internal comparison of costs.

In any company (and Ferodo is no exception) it is likely that the costs of some functions or activities will change out of line with other costs. This is often true when comparing with each other the costs of, say:

   direct labour
   indirect labour

factory or general works overheads
sales, distribution and administration.

There may be some very good reason why direct labour costs have increased when compared with general factory labour costs, but they have to be known and understood. Any discrepancy which appears to be out of line should be challenged.

1.3 CHANGING COURSE

No one should consider a manufacturing strategy without first reviewing the company's past performance and its current product market position. Both of these aspects will be considered in detail later. Tough cost reduction and cost control might save a company which is about to go over the cliff edge, but quite obviously something more sophisticated is needed to ensure long-term prosperity. Many companies now issue a 'Corporate Mission Statement' which lists the philosophical characteristics needed if the company is to be a viable long-term unit. Some of these statements are very general:

Be capable of competing on equal terms with any other similar manufacturer in the world.
Achieve high and consistent productivity.
Raise productivity by at least 5% on a year-by-year basis.
Provide a level of service, product quality and price which achieves a market share of at least 30%.

The Ford Motor Company's statement records the following:

'Ford is a worldwide leader in automotive and automotive-related products and services as well as in nearer industries such as aerospace, communications and financial services. Our mission is to improve continually our products and services to meet customer needs, allowing us to prosper as a business and to provide a reasonable return for our stockholders, the owners to the business'.

The statement then goes on to record various values (people, products and profits) and guiding principles (quality, continuous improvement, employee involvement, dealers and integrity).

Some are much more prosaic in setting down a considered view of 'What kind of company do we have to be?' to be a viable and prosperous unit. On all possible occasions the various factors need to be re-stated and if necessary adapted—a little.

My personal input to our own early debate at Ferodo was as follows (the opinions expressed were not accepted by everybody and in total were never Ferodo strategy):

'It seems impossible in the motor components industry to achieve price rises remotely near to the pay rises and general cost increases which everyone in the company thinks are vital. If productivity improvements do not fill the gap then it is quite easy to forecast when we will go out of business. One myth which must be killed is that it is impossible to carry out production in the UK in the way it is done in Japan. The people who now work in Nissan, Toshiba and Hitachi (and soon no doubt Toyota) in this country have proved that a UK workforce can be just as efficient as its Japanese counterpart. The problem for most established UK companies is that they do not start from a green-field site, with a single-union agreement and with a potential to create a culture appropriate to being a world-class manufacturer. This is not to say that it is impossible to create a green-field mentality in a company which has been on one site for fifty years, it is just a little more difficult. So, if competition is getting tougher all the time, and price rises of any substance difficult to achieve, how can we create a brown-field if not a green-field culture? Perhaps by dealing with these factors in the proposed way'.

1.3.1 People
Perhaps the most fundamental way in which Japanese companies differ from British organizations lies in the way people are treated. A major Japanese industrialist (Konosuke Matsushita) put it this way:

'We are going to continue to improve and the industrial West is going to lose out. There is nothing you can do about this because the reasons for your failure are in yourselves. With your bosses doing the thinking while the shop-floor workers wield the screwdriver, you are convinced deep down that this is the right way to run a business'.

In the UK the belief that the management cadre is the only body of people capable of intelligent thought dies hard. It is perpetuated by our educational system and by the prevalent general view on class and worth. Even when the most senior manager in the organization decrees that a 'change in culture' is vital for the future of the company, his senior manager through inertia or downright opposition can destroy the approach. Let no one in the UK underestimate the power of established privilege in a strongly class-ridden society. The senior executive has got to be exceptionally strong-minded to overcome a lifetime's conditioning in the minds of his subordinates.

If a company cannot exist without people, then the people equally need the company. The relationship should be symbiotic. Good pay and good conditions are perhaps only a minor part in the establishment of a cultural situation which makes sense in the 1990s.

1.3.2 Adaptability
The only consistent factor in the future of a UK manufacturing company will be the need for change. Product markets, technologies, relationships, systems are all changing and will continue to change. Equally, societies are changing. Social and economic pressures in the UK are now totally different from what they were in the seventies. The companies that failed to adapt then are no longer around now. Adaptability and change must become a way of life   in jobs, in how we pay people, in organization, in

technologies and in how things are done generally. Nothing should be treated as if it had any permanence any more.

If a key customer changes his mind about his scheduled requirements then he must be accommodated quickly and his new requirements must not be regarded as an impossible imposition. If new environmental legislation is introduced then it must be put into practice in the shortest possible time.

Investment must be viewed as a comparatively short-term application of self-generated funds. Most British companies have grown up using depreciation rates of 7% or even 5%. This cannot be acceptable and we need to move rapidly to rates of 15% or higher even for the most mundane equipment, with 33% at least for high technology applications—despite what this might do for short-term profit.

1.3.3 Profits and costs
Without reasonable and consistent profits a company is doomed. Profit needs to be earned consistently and at an appropriate level to provide shareholders with a reasonable return. Then we can all get on with the running of the company without the need to look over our shoulders constantly.

When a profit plan is made, everyone should know what the key objectives or targets are and how they are to be achieved. Once the plan is in operation it needs to be monitored carefully. The profit plan should be a mechanism to drive the company to success. Its achievement or not should be the main determining factor in whether senior managers keep their jobs. It should eventually be a major element in their pay awards.

1.3.4 Customer and sales
If the right products are being made, then there are three other key factors which produce sales—prices, quality and service. Getting any one of these wrong loses sales revenue. Successful companies handle their customers well. They set out to supply, support and nurture them. They achieve a special relationship with them.

Quality is very much a manufacturing problem. Customers are demanding that 'zero-defects' is essential. This might sound utopian, but it is the wave of the future. Ignore it and customers will just disappear.

1.3.5 The shop floor
Even in a manufacturing company the number of people who want to be closely involved with production is limited. Sales/marketing often appears a more attractive and highly paid occupation. Sitting in an aeroplane or driving around in a company car seems to have a tremendous fascination for many people. Yet how efficiently raw materials, bought-out components and services are converted into finished products largely determines the profitability of the unit. Manufacturing and its ancillary service such as engineering and maintenance normally account for up to 70% of the total costs of a manufacturing company.

The shop floor, with its possibilities for improving productivity and quality and reducing work in progress while improving customer services at the same time, should loom large in the allocating of company resources. A significant part of the organization's talent, its computer resources, its systems development and general investment should be directed towards the shop floor. The rewards for intelligent people to work there should be among the highest in the company.

1.3.6 The technological base of the company
To offset the labour cost advantages which most Second and Third World countries have, manufacturing companies must obtain the best technology available and use it effectively.

If our products are to be of an ever-increasing quality and to give a technically superior performance, then there is a need to put more money into R&D than has been done before. As a proportion of revenue, the West Germans spend anything up to double our own R&D spending.

Technology should be used to help improve quality and productivity and also to speed up the whole process of making and despatching products. The humdrum, manual jobs where operatives repeat one activity for the whole of a shift should practically all be eliminated. Technology applications may actually reduce labour to a degree, but without them there may be no jobs at all. So far, technological applications may actually have increased rather then reduced numbers employed.

1.3.7 Simplification
Complexity adds to confusion. It increases bureaucracy, it builds up costs, it debilitates efficiency. If everything was kept simple then the effect would be the reverse of that. It seems in the nature of things that complexity usually triumphs. It is all too easy to add to a system, a technology, an organization or a routine.

Simplification must take place in the:

- product range
- payment systems
- order processing
- organization
- jobs
- methods of reporting.

1.3.8 Cash and working capital
Companies go broke by not making profit. They go broke faster still if they do not generate cash. There have been times in our history when stock was made even though it was not immediately needed. Indeed, occasionally overtime has been worked to make products which have been extremely slow to sell.

Stock in all its forms—raw materials, work-in-progress and finished goods—has to be reduced to an absolute minimum. Whether it is possible to operate the company efficiently on a 'just-in-time' system is still to be fairly tested. There are chances that costs may rise.

1.3.9 Reductions in non-manufacturing costs

Non-manufacturing costs are those which are largely contained in the sales, administration and distribution activities. One of the most important ways in which a company can improve its flexibility and fitness in the face of changing environments, is to have a low break-even point. By that is meant that we need to be able to make profit at the lowest possible utilization of our installed capacity. Companies which do not make profit until, say, 90% of installed capacity is being used, leave themselves wide open to changes in sales patterns and markets. The lower the break-even point the better. Hence the need to reduce SDA costs and keep them at a level consistent with a low break-even point.

1.3.10 Rewards and punishments

Four clear-cut, unambiguous criteria seem relevant under this heading.

(a) A company should pay its people what it can afford. If it pays more than that, then something else will suffer.
(b) Individuals should be paid for their part in achieving overall company performance. For some people, effort may be a key element, but all of us should know that what is done and how well it is done reflects on company performance and should be paid accordingly—whether this is in achieving good output, reducing rejects or saving on energy costs. A good performance is not just one which ensures that people are paid well at the end of the week or month, but one which ensures that there is a job to go to in two or three years' time. Only good overall company performance can ensure that.
(c) If productivity in the company does go up, then the people responsible should be able to expect that their performance will be rewarded commensurately. If productivity goes down, then some pain in relationship to that decline is equally essential.
(d) No one can work in isolation from other people. Everyone depends to some degree on colleagues, Even the man working by himself on a machine depends on people in previous operations to give him the right quality of work to operate on.
    As more new equipment is put into a factory, more flexibility is demanded by customers, and as performance generally needs to improve to compensate for low product price rises, then small groups of people, instead of individuals, grow more important. So group pay schemes rather than individual payments are essential.
(e) There should be a downside to rewards as well as an upside. Inadequate performance should be punished in the same way that good performance is rewarded. If everyone agrees to have improved status and rewards when things go well, then there is nothing wrong in having some pain for poor achievement.

1.3.11 Conclusion

The amount of latent cynicism in a British manufacturing company should never be underestimated. While everyone agrees that change of some sort is necessary, what change and how quickly it should be introduced varies considerably in people's opinion. It depends to some extent on whether the people concerned work on the shop floor, or in offices, are managers or shop stewards, belong to the T&GWU or the CSEU, are technically qualified or not, work in production or sales.

People will remain cynical until they see that changes are being made and that senior managers are firmly committed to them. Commitment must be demonstrated on every day of the week and at weekends if overtime is being worked.

So, how, to convert a fairly pious 'What kind of company do we need to be?' into reality? Once appropriate environmental analysis has been carried out the following proposals show how it might be done.

1.4 THE POSITION PAPER

It may seem pedantic to state that as well as a 'What kind of a company do we need to be?' record, a position paper is needed; but there are differences between the two. The position paper statement should say fairly briefly what relationships with its customers, markets and suppliers the company needs if it is to prosper, and what in consequence might be its internal position. An appropriate position paper might be:

> We need to become one of the top players in our business.
> Only by being a senior player will we have sufficient leverage to gain the kind of orders which are essential to keep installed capacity in full operation.
> While we have made significant steps to re-equip our manufacturing processes with an increasing R&D budget, we need to match competitors in our key markets.
> We must operate on a worldwide basis and believe that 'Taking on the World' is a key philosophy within our organization.
> We must equal the best of our competitors in terms of service, quality and product costs.
> Considerable training has been done but more needs to be done to ensure that our people handle MRP II, TQM, JIT effectively.
> Even so, we will fail unless there is a significant change in company culture, reflected in communications, work organization, job responsibilities, agreed objective setting and above all, how people are paid.

1.5 THE PRODUCTION FRAMEWORK

The bitter lesson we learned in Ferodo, and which probably other manufacturing units came to understand as well, is that if more than marginal improvements are needed, then it is necessary to analyse and act upon the total manufacturing system. Even on a slightly less fundamental basis, the relationship between resource use and output needs to be established. Goal conflict has to be reconciled. The total system needs to be adaptive to the pressures put upon it, whether these are external or internal.

It has been a particularly British trait to embrace some technique or procedure which apparently offers to solve all immediate and future problems. There have been countless examples—management by objectives, job enrichment, operations research, computers, even, once upon a time, work study. Most senior production managers have been guilty of rushing in and applying one or other of these techniques, only to retreat, slightly baffled, because whatever improvement had occurred was very small, and on some occasions the whole exercise had been counter-productive.

Obviously a wider view is needed. If the production system as a whole is to be improved then the whole framework of production needs to be considered.

What were we looking for? What was needed was an organizational framework which would cross functional boundaries and at the same time interrelate one with another. The benefits of any such framework, we thought, were as follows:

(a) Interrelations of all kinds can be seen and put into the profit plan.
(b) Functional boundaries, which have often stifled performance improvement, could be crossed easily.
(c) Opportunities for improvement, which in traditional methods of analysis and planning might have been missed, are recognized.
(d) Corporate rather than individual objectives can be pursued.
(e) The mistake of concentrating on one technique or activity is eliminated.
(f) Management at all levels has to think in a cohesive, corporate way.
(g) The organization can act as one entity, not a series of unconnected activities.

(Such a framework can be used by any organization, perhaps not necessarily carrying on manufacturing activities.)

So, what is the framework? It is a series of interrelated components which together make up the whole process, and is set out in Fig. 1.2.

(1) The product market
(2) Technology
(3) Resources
(4) Work organization and training
(5) Systems and information technology
(6) Motivation and reward systems.

Why this framework?

(1) *The product market*
Failure in the market place is often reported as a major factor in corporate extinction. It is not just product design and associated marketing which could fail but product costs, quality and service, which would normally be related to the production process. Marketing starts on the shop floor.

(2) *Technology*
Technology could be the key factor in improving organizational performance, but technological improvements need to be funded and produce an adequate return. This is unlikely to happen unless there is a well-trained and dedicated workforce which will support new technology.

	WORK ORGANIZATION AND TRAINING	
PRODUCT MARKET	Structure Job design	RESOURCES

Strategic fit
Product range
Globalization
Quality
TQM
	Teams
	Solid technical systems
	Training purpose and effect
		Working capital:
		    Cash
		    Debtors
		    Stock
		Fixed capital:
		    Machinery
		    Buildings
		People
		Material

TECHNOLOGY	THE PRODUCTION SYSTEM	
		MOTIVATION AND REWARD SYSTEMS
World-class manufacture FMS Robotics CAD/CAM OPT JIT	SYSTEMS AND IT	Traditional PBR Multi-factor incentive Second-generation incentives
	Planning of the organization MRP II IT Management Accounting	

Fig. 1.2. The framework and some of its components

(3) *Resources*
Good organizations minimize the use of resources and maximize their benefits. They employ the minimum of well-trained people. They maximize the use of materials and energy. They keep stocks to the lowest possible level. Cash flow is regarded as very important.

(4) *Work organization and training*
A fully effective organization is one where company objectives are accepted by everyone. Teams or other forms of organization are established so that all concerned contribute fully (within their authority and responsibilities) to achieve objectives.

(5) *Systems and information technology*
There are two major systems which all organizations need to operate effectively — the order processing/production planning system and the profit planning/resource control activity (e.g. MRP II and Management Accounting).

(6) *Motivation and reward systems*
What people in the manufacturing organization can be paid will largely depend upon their productivity and the company's ability to contain cost increases to levels which do not make their products over-priced. Achieving a satisfactory relationship between inflation, competition and productivity is crucial.

Element	Key objective	Methodology
(1) Product/market	Quality, service and price equal to competitors now and for the foreseeable future	Market segmentation. Added value factors applied to the most basic products. Product range appropriate for maximizing contribution but no more. Differentiation in: – products – markets – market segment Total quality management
(2) Technologies	To have production technology and R&D effort equal, at least, to competitors and, wherever possible, superior	R&D product/market driven where achieving contribution is important. Capital expenditure directed towards production facilities which are computer planned and controlled, to use minimum labour and achieve flexibility in making the product range. Ensure high quality to predetermined standards continually. Robotics
(3) Resources	Minimize total resources used. Maximize their use, especially machines and raw materials. Own as little as possible. Depreciate at the highest possible rate consistent with changes in technologies and making profit.	Identify resources: – people – space/buildings – production – machinery – materials – other working capital and relate to forecast. Use Just-in-Time and Flexible Manufacturing Systems to minimize space and equipment needed. Maximize material usage by material-utilization and reject-elimination studies. Working capital control, stock, debtors, etc
(4) Work organization	The organizational structures should help to create a culture and power/authority/responsibility relationships which ensure everyone in the organization understands corporate strategies, accepts associated objectives, and is told of results so that they will know what to do to improve them	Resolution of goal conflicts. Group structures. Minimization of hierarchies. Socio-technical organization design. Promotion of 'leaders' irrespective of qualifications. Training in objective setting and achievement. Job design. Communications
(5) Systems	Systems should ensure that resources of all kinds are planned, utilized and controlled, to achieve corporate objectives	Materials Requirements Planning II Contribution Costing. Measurement of key ratios. Measurement of specific targets for all personnel: – revenue – contribution – added value. Measurement of relationships between the use of resources and what is achieved
(6) Motivation and reward systems	Rewards should be related to corporate success, group and individual achievement of part of that success, not just individual effort	Payment related to profit, added value and contribution achieved. Second-generation incentives

Fig. 1.3 The manufacturing strategy table

Figs 1.2 and 1.3 show the production framework, objectives and methodology.

The uncompromising belief being put forward is that one component in the framework cannot have an independent existence. It cannot be regarded as something separate and standing alone. Much writing on management has postulated the opposite and has been very wrong in doing so.

The simple relationship chart shown as Fig. 1.4 tends to prove the idea. For example, if a robotic manufacturing cell is introduced, it is very likely that team working, the local payment system and shop floor controls will all need to be changed if full benefits are to be achieved from the investment.

Good manufacturing management will be aware of most of the current best practices within the framework. Whether they actually apply them immediately should be debated. It is probable, for example, that new technology is needed urgently and this by all traditional analysis should be pressed vigorously. Yet very often, despite prodding from the DTI and others, it is not.

The euphoria of the 1980s has seen many companies over-invested. When interest rates rose they had too little net profit to pay back money borrowed which funded their new investment.

The failure of many UK manufacturing companies to equal their foreign rivals has brought about something of an inferiority complex in them. Management is despatched to Japan or the Nissan plant in Sunderland in the hope that somehow, somewhere, they will find the holy grail of constantly improving efficiency and productivity, zero defects, total quality management and so on.

All these may be prized goals, but how they are achieved could be quite different in Osaka and in Manchester. No two countries or even companies are alike. Always be wary of the recent visitors to another company or someone who has just read an article on 'The best factories in Britain'. The production framework, from experience, is the best possible basis for setting out to improve manufacturing performance. The road for each company could be different. History, culture, the current competence of management will all dictate the way and the pace of change. There is no single ready-made plan.

There are, though, some common threads which all companies might consider; for example, some aspects of past management which are unlikely to prove of value in the future, such as:

(a) F. W. Taylor-type incentives.
(b) Continuation of organizational hierarchies with assured status.
(c) An endlessly extended product range 'because this is what our customers want.'
(d) The payment of a going rate to all employees, irrespective of whether the company can afford it or not,
(e) Allowing customers to add value to products made by the company.
(f) Believing that doing things in-house, with company staff, is always better and inevitably cheaper than obtaining services or even products from outside.
(g) To this list might be added the failure to apply such well established activities as:
  cash control
  inventory control
  management accounting.

16 The manufacturing framework [Ch. 1

	Product market	Technology	Work organization and training	Resources	Systems and information technology	Motivation and reward systems
Product market		To help achieve strategic fit	To give best possible customer service		Marketing data	
Technology	To gain strategic fit			To make effective use of new investment		To ensure new investment is met effectively
Work organization and training	To improve service	Training needed in new equipment				
Resources		Cash generated for investment				
Systems and IT	Market knowledge and operational planning			Systems development directed towards minimum resource use for maximum output	Planning and control of resource usage	
Motivation and reward systems	To achieve service on time					

Fig. 1.4. Relationships in the production framework.

Sec. 1.5] The production framework 17

The analysis can be lengthy and complex. The following should be included:

(1) Policy/strategy
    Financial pressures
    Management competence
    Competitive pressures
    City pressures
    Labour history
(2) Product market
    Production process and its relationship to the product market
    Competition
    Strategic Fit
(3) Technology
    Opportunities
    Cost/financing
    Space/factory layout
    Relationships with training and work organization
(4) Resources
    What resources? (everything that incurs a cost)
    Labour and culture
    Material—control and quality
    Stock and JIT/zero defects
(5) Work organization and training
    Culture of the company
    Hierarchies
    Function and failings
    Roles and objectives
    Leadership
    Training – specialist skills
        – create corporate cohesiveness
        – ensure profitable change
        – work organizations to operate effectively
(6) Systems and information technology
    User specifications
    MRP II
    Management Accounting
(7) Motivation and reward systems
    Productivity and pay
    Current wage rates
    Systems of reward
    Pay policy
(8) General

Handling change—effectiveness
Trends in productivity
Failures and strengths—reasons:
Culture
Training
Leadership
Communication
Strategy assessment.

As a concomitant of this analysis, objectives need to be set.

1.6 SETTING KEY OBJECTIVES

It is possible that, as the Japanese show, the short-term pursuit of financial gain may not always be conducive to achieving long-term success. Equally, the failure of some UK companies to achieve a satisfactory profit, in the long term, has had disastrous consequences.

Without hard objectives, companies end up lacking dynamic and urgency. They will fail.

Books on profit planning[†] normally pay great attention to 'gap analysis'—what the company is achieving now in the way of profit or cash and what it could or should do. Profit planning was at one time not welcome in T&N, because, apparently, it gave managers an easy time, achieving easy objectives. Objectives need to be tough, comparable with the best being achieved in the industry. If one motor components company consistently achieved a return on capital employed of 32% or more, why should other companies in the same industry struggle to reach 20%? There must be something wrong in the way the latter companies are being managed.

So, what objectives and what kind of returns? Probably the following:

Return on average capital employed   25% seems to be the least that should be targeted. If the cost of borrowing money is 12% or more, then the risk element dictates this kind of return.
Operating profit/turnover   Some good UK manufacturing companies achieve better than 14%. So 12% is a minimum requirement.
Turnover/Capital Employed   2.4 is about right.
Stock ratios (raw materials, components, work-in-progress, finished goods)   These depend, to some extent, on the industry, product range, possibility of having consignment stocks, importance given to reducing WIP as opposed to minimizing set-ups, etc.; but 80 days in total seems a maximum figure.
Debtors (export/home)   50 days, depending on balance between home and export sales.
Net working capital/turnover   14% could be a satisfactory figure.
Payroll costs/added value   Any figure greater than 75% should be highly suspect.

[†] B. H. Walley (1978) *Profit Planning Handbook*. Business Books.

Turnover per employee > £35 000 (not a ratio on which too much weight should be placed.)

Profit per employee £5000 + is probably correct if average pay is approximately £12 000 per person.

These are all figures that a reasonably efficient, well-managed UK company should achieve. As such, they should be used as a starting point for gap analysis—what is currently being achieved and what might be a reasonable target.

1.7 PROFIT PLANNING IN A MANUFACTURING COMPANY

Profit planning has often been dominated by accountants and, perhaps less now than previously, by professional planners. This is wrong. Profit planning is everybody's concern. Shop stewards, junior managers, specialists of all kinds, should be involved in the construction of the plan just as much as senior management must be.

Using the production framework as the basis of the profit plan, produces a different set of information and plan formats than in an accountancy-dominated planning activity. This helps to prove the point that using the framework tends to make planning and operation far more realistic.

The elements of the plan should be these:

(a) *Top-down planning*
A key factor in planning should be the setting of key objectives by the Chief Executive or the Board of directors. Whether these are the figures set out previously or some other criteria matters less than the starting point. The plan philosophy should be a 'top-down' one. Middle managers should be informed of their objectives and asked to stipulate actions which will achieve them. Traditionally the converse has been true: data is aggregated, and sales forecasts, costs, inflation rates, etc. and a planned profit eventually emerges. Little wonder that most plans in the past have had little 'stretch' in them.

(b) *Profit and Loss account plan sheet* (Fig. 1.5) The top-down plan requirement, agreed by the Board, is quoted on the sheet in column 'B'. The current year's performance and the plan for the next period or year is then compared with the top-down requirement and the differences quoted under the heading 'Gap'.

The debate surrounding the plan should then be mainly concerned with improving on the current year's performance and seeing how much further the 'Gaps' which have been highlighted can be closed.

The Profit and Loss account sheet shows the conventional information which leads to profit before tax and cash generation.

(c) *Balance sheet and key ratios* (Fig. 1.6)
This sheet should also show the top-down requirement, but recorded against the balance sheet items and key ratios. These latter show the base data set out in a different form.

20 The manufacturing framework [Ch. 1

Element	Current and year's performance A	Top-down plan requirement B	Plan for next year C	Gap A–B	Gap B–C
Net sale factory cost					
Gross margin					
SDA costs					
Operating profit					
Finance costs					
Exceptional items					
Profit before tax					
Depreciation					
Fixed asset sales					
Working capital movements					
Cash generation					
Cash utilized (CAPEX)					
Cash surplus					

Fig. 1.5. Profit and Loss account plan sheet.

	Current performance A	Standard B	Planned performance C	Gap A–B	Gap C–B
Working capital					
Raw materials					
WIP					
Finished goods stocks					
—debtors					
Creditors and accruals					
—provisions					
Other liabilities					
Net working capital					
Fixed capital					
—land					
—buildings					
Plant and machinery					
Total capital					
Return on capital employed					
Operating profit/TO					
TO/capital employed					
Added value/TO					
Payroll costs/av.					
TO per employee					
Operating profit per employee					

Fig. 1.6. Balance sheet and key ratios.

Sec. 1.7] Profit planning in a manufacturing company 21

(d) *Cash flow* (Fig 1.7)
This part of the plan should show the cash to be generated each month, during the planning period. It is possible that some months may see a negative cash flow and the method of funding this should be pre-determined.

	Jan.	Feb.	March	April	May	June	July	Aug.	Sept.	Oct.	Nov.	Dec.
Cash generation												
Operating profit												
Dividends received												
Depreciation												
Fixed capital disposals												
Working capital												
Provision movements												
Cash generated												
Stocks												
Debtors												
Creditors and accruals												
Working capital movements												
Capital expenditure												
Interest payable												
Tax												
Dividend paid												
Net cash flow funded by:												
Bank borrowing												
Bank debtors												
Overdraft												
Other												

Fig. 1.7. Cash generation plan.

(e) *Framework details* (Fig. 1.8)
The cash flow plan sheet should probably end the accountancy section of the plan. The remaining sheets should set out what planned activities and changes are to be carried out in each of the elements of the framework. Fig. 1.8 should summarize the details as follows:

1. Key objectives: These should already have been recorded in the corporate mission statement.
2. Best practice: An indication is needed that the 'best practice' in the appropriate element has been determined or at least considered. The 'best practice' could be what is taking place in Japan, the USA or Germany, or even be some theoretical proposal, put forward by an industrial academic. Whether the 'best practice' is appropriate for the company should be stated and reasons for the decisions recorded.

22 The manufacturing framework   [Ch. 1

	Key objectives	Best practice known and understood	Relationships	Planned effect on P & L account balance sheet and cash
Product market	Strategic fit TQM			
Technology	World-class manufacturer			
Resources	Maximize cash flow			
Work organization and training	Team activities throughout the organization			
System and IT	Effective MRP II and management accounting			
Motivation and Reward systems	End of PBR Harmonization of conditions			

Fig. 1.8. Framework plan.

3. Relationships between the elements in the framework need to be carefully considered and recorded, with the belief that altering one part of the framework will obviously have some effect on the remainder.
4. The final column should show what the effect will be if the item plan is achieved successfully. The changes in the Profit and Loss sheet, the Balance sheet proforma or the Cash performance sheet should be known.

Individual plans for each framework element are then needed.

1.7.1 Target achievement

In my hearing, the greatest criticism that Lord Tombs made of Ferodo was that in the past (prior to 1982) Ferodo rarely, if ever, hit a target it had put into its profit plans. There seemed to be some truth in this comment.
   The profit planning process should:

- record agreed objectives
- agree suitable strategies

- construct appropriate action plans
- ensure that they are achieved.

To achieve these objectives, chop, change, drive, threaten! No one should get away with not achieving an agreed objective, unless the surrounding circumstances could not have been foreseen; and this should not be a common occurrence.

1.8 THE ROLE OF MANAGEMENT, ESPECIALLY THE SENIOR MANAGER IN THE ORGANIZATION

The longer my work experience has gone on, the more important the role of the most senior manager in an organization appears to me to be. The Japanese may believe that they run their organizations with consensus management; it is rarely so in the UK.

The need to change may be apparent among middle management; perhaps even the shop floor are aware of it: but never in my experience has it been possible to generate fundamental change solely from the middle ranks of an organization. Without the ambition and drive of senior management, the careful planning of a manufacturing strategy can be a waste of time. Hence, the activities of the senior manager must follow these lines:

(a) He must be seen and known. Walking round the factory every day and talking to as many people as possible always seemed important to me.
(b) He must give a lead both in his patterns of work and in his general behaviour.
(c) He must promote and reward the right people. Some Boards of Directors, in my experience, tend to perpetuate themselves, promoting similar people, with similar views and similar failings. It takes a strong-minded Managing Director to promote a maverick, even though doing this could be the saving of the organization.
(d) He must establish a well-defined method, set of procedures, philosophy or system for running the company. In Ferodo we talked about the 'train principle'. It did not matter if some people were less competent than others, if they were linked up to a train where the competent people drove the engine. Then everyone went in the same direction, whether they wanted to or not. My national service taught me very little except that it is possible to run a relatively efficient unit with mediocre officers, by having a 'system' of disciplines and procedures which everyone understands and to a degree, follows. Companies even with a large number of poorish managers can be run in the same way and achieve reasonable results.
(e) He must train. When Wellington called his troops 'the scum of the earth', he quickly added, 'It is only wonderful that we should be able to make so much out of them.' He won battles by rigorous discipline and drill. All sensible and successful companies operate in the same way. There is a company system. People are trained to understand it, to use it and to refer to it on all occasions.
(f) He must not be afraid to have few friends or to be disliked. Friendship is always difficult when discipline has to be enforced.
(g) He must communicate. He must tell his people what is going on, what the company plans are, what the results are. He must indicate what corrective action is needed

if plans are not being achieved, and what good or bad performance individuals and departments and even managers achieved. He must tell people what changes are being made, especially the people who will be directly affected. He must be able to talk without condescension to shop-floor operatives and stewards. He must be patient.

(h) He must be capable of attracting trust. People must believe what he says. Equally he must trust his subordinates, but not too far. He must delegate, but never so much that the whole company plan goes wrong because of the delegation.

(i) He must always try to give the people under his control the best possible equipment, environment and personal conditions. They must know that he will always do his best for them. They should believe that he will pay them as well as the company can afford and will not sacrifice their jobs or allow any deterioration in their working conditions, unless it is absolutely unavoidable.

(j) His role is in the company, not outside it. He must not carry out extraneous and unnecessary duties. It is far too easy to join trade associations or a local quango, and avoid tedious work back at base. Following extraneous activities is rarely useful.

(k) Subordinates must know that if they fail in their jobs without good cause, retribution will inevitably follow. The senior manager must be capable of rewarding and punishing with equal facility.

1.9 CONCLUSION

It often seems true in the UK that only the second-rate or the eccentric bother to work in manufacturing (though most of us in manufacturing would strongly disagree). No matter what occurs in the next few years by way of training, recruitment or even improving liaison between schools and industry, there will be a continuing dearth of highly intelligent, highly motivated potential senior managers.

The *Financial Times* reported on 14 July 1989 that students still shun industry as a career. Only 7% had thought seriously of joining a manufacturing organization.

The situation in the USA appears nearly as bleak. *Business Week*, on 19 September 1988, published a special report on '*Human Capital*' *with the conclusion*:

'The nation's ability to compete is threatened by inadequate investment in our most important resource—people. Put simply, too many workers lack the skills to perform more demanding jobs.'

How do the rest of the Common Market countries and perhaps some countries in Asia and Africa compare?

In the *Financial Times* of 26 June 1985, I wrote an article[†] in which I related my experiences of being seconded to work in (then) West Germany. Even the efficient Germans had made a mess of running 'BERAL', a company T&N had taken over

---

[†]'Report from the Front Line—confidence gets a boost.'

after it had gone bankrupt. The Germans, though, do take industry seriously, as, in my experience, do the Italians. German managers are highly trained—and highly paid. My first crisis meeting in BERAL was attended by eight German managers, all of whom were qualified engineers. The threat of closure of the plant soon brought the local *Bürgermeister* round in his BMW to see what the town and its council could do to help.

Engineering, rather than basic management, is well regarded in Germany. For a nation which gave the world Kant, Hegel and Marx, management philosophy is not strong.

The UK companies which have recently done well—BTR, Hanson Trust, GECo (for a while)—have largely relied on tough management accounting controls and not engineering innovation as such. '*Vorsprung durch technik*' (as Audi used to say) may still win export markets that the UK has forgotten about. On the other hand, the hard uncompromising approach to containing cost of all kinds that we practised was missing in Germany. The need to achieve a high net margin or return on investment did not rate too highly. Business skills seemed subordinate to engineering excellence.

What both countries need is a marriage of technical ability with rigorous financial control. The accountancy-trained engineer, or the accountant with a strong knowledge of engineering, should be the key type of person in the company.

All my experience of the UK, Germany, Italy, India and Nigeria suggests that:

(1) Every company has a different environment in which to work, its culture is different, its history is different.
(2) Even if the same products are being made with nearly the same machinery, using the same techniques or systems could be wrong.
(3) What is needed is a framework which can be used to develop local, carefully structured plans. The manufacturing framework has universal application. What results from its use could be dissimilar from one country to another.

CONTENTION

(1) Major profitable change in a manufacturing company can only be achieved by adopting a total manufacturing strategy, based on a well-considered framework. Piecemeal/*ad hoc* changes, no matter how hard they are pushed, will fail.
(2) The method which should be followed is:
   (a) Carry out environmental analysis.
   (b) Designate the kind of company needed to achieve long-term profitability, within the determined environment.
   (c) Agree a framework, around which a strategy can be developed.
   (d) Agree objectives and appropriate gap analysis.
   (e) Produce action plans which can be monitored successfully.
   (f) Reward and punish according to achievement.
(3) Success in a manufacturing company will only be achieved from within. The senior management must want to succeed and thoroughly understand what must be

done to gain success. It is not the government, or society, or cheap imports, or even trade unions which has been the major cause of industrial decline in the UK: it is management.

A suitable set of strategies
In helping the company to become a world-class manufacturer we will need to:
(1) Develop an integrated manufacturing strategy based on a declared framework.
(2) Develop competitive marketing strategies which recognize, by product market or market segment, the competence profile of other suppliers/competition. Then using this analysis initiate action, particularly resource deployment which will gain competitive edge.
(3) Ensure that through local structure/organization, culture, pay and motivation, all people make the maximum contribution towards achieving world-class manufacturing status.
(4) Ensure that all our manufacturing technology is as good or better than that of our competitors. A key way in which we can compete against low cost imports from the Third World is to deploy better manufacturing technology then they can and use it more effectively. It should be deployed where a return of 35% or more is achievable. It should preferably be computer-controlled and -designed to eliminate as much manual intervention as possible.
(5) Offset the low price rises we can get for our products by ensuring that year-by-year productivity improvements more than regain the difference between prices and potential costs.
(6) Use contribution margins as a means of deploying resources in the short term, but in the long term use resources to ensure long-term viability and high profitability combined.
(7) Reduce variable and fixed costs further to bring back our break-even point.
(8) Introduce organization changes which make effective team units with appropriate autonomy for local resource utilization.
(9) Enhance the planning and control of the company through the application of MRP II and an improved management accounting system.
(10) Eliminate as far as possible a curse of British industry—relative deprivation—by company-wide job evaluation, single status and pay-related input to company performance.

2
The product market

2.1 INTRODUCTION

Failure in the market is a key factor in corporate extinction. The British motor components industry suffered because of the decline in national car assembly. One major reason for this was the way in which the old British Leyland organization got into trouble. It produced a series of cars which few people wanted to buy. The Allegro was much inferior to the VW Golf. The Ambassador Princess was no match for the Ford Cortina.

The Marina was one of the worst of its class. In all cases, performance, style and, above all, quality were lacking. No matter if BL's internal industrial relations were poor, the main failing was in the market place.

Yet a company dominated by salesmen, with information systems designed by marketing people and the factory turned upside down every day by unthinking product designers, could also be led to disaster.

Ferodo once had a product range numerically greater than the number of car types on the road. No attempt had ever been made to measure the importance of one product compared with another. In the eyes of salesmen, all were of equal value. A simple Pareto analysis would have proved that most revenue came, as usual, from a very small part of the range.

A similar situation produced by a lack of commercial discipline was discovered in BERAL, the Ferodo equivalent in West Germany. In this factory there was shelf after shelf of dusty, rusting and expensive tools. Some had only been used once and then forgotten—a legacy of the salesmen's demand that they must have every product available that their customers were likely to want. Like everyone else, sales people need to accept financial discipline and must learn to behave responsibly.

2.2 PRESSURES TO CHANGE

Are things changing and, if so, how?

2.2.1 A business guru's change predictions

American business evangelist Tom Peters[†], has suggested that the 1990s will show the following:

(a) There are no mature products—everything can be changed in some way.
(b) Niche marketing—the favourite marketing ploy of the eighties—will need to give way to one-to-one relationships, where companies relate their products to single customers. The future will be concerned with manic specialization.
(c) Information about customer preferences will become crucial and probably more important even than manufacturing the products concerned.
(d) Smartness in products must be equalled in design, service, marketing and distribution.
 (All of which could be re-run of the old ideas of identifying potential customers and then adapting products and associated services to their requirements. Or conversely, making sure that the potential customer was aware of what you had to offer.)
(e) Market segmentation. One key factor which marketing people thought was essential to effective selling was market segmentation. This was defined as 'part of an overall market, which would respond to one set of marketing stimuli'. It was a separate and distinct portion of a total market. By segmenting markets, it was hoped that the marketing thrust would be directed and pointed. General approaches were eschewed.

What Tom Peters is saying is that marketing segmentation may end up as a single customer buying a single product. This may have some validity. Buying has always been a personal individual choice—about colour and fitting, in some cases. But selling disc brake pads needs little concern for the individual buying the car which they fit. The guru is not always right. Expectations could prove far greater than the rule.

2.2.2 An evaluator report to the DTI

A more prosaic approach to the changes which could be pressuring product markets in the 1990s resulted from a report commissioned from PA Consultants by the Department of Trade and Industry, whose findings were published in the *Financial Times* of 29 November 1989. They provide interesting reading for anyone involved in looking at product markets with an overall manufacturing strategy[‡].

(a) *Global economy*   Manufacturers, it was said, must increasingly think of the world as one market in both the selling and buying of raw materials and components. Ferodo's own strategy certainly followed this line, but whether selling in Khatmandu really made sense is debateable.

[†] Tom Peters. Conference jointly organized with *The Economist*, 13 February 1990.
[‡] *Manufacturing into the late 1990s* (1989) HMSO. ISBN 0–11–5152067 £20.

(b) *Demographic trends*   Throughout most of the Western world and Japan, there are ageing populations with changing product demands.
(c) *The environment*   The 'Green' lobby is growing more and more important and manufacturers should make environmentally friendly products in a responsible way. Energy-efficient production is another trend, though anyone faced with rising UK electricity prices will already have done all possible to conserve it.
(d) *High added value products*   To compete with comparatively cheap products made with cheap labour, the West needs to concentrate on high added value products.
(e) *Product proliferation*   Potential buyers are no longer happy to have a basic product; they would like product differentiation. Diverse product manufacture could become a norm and flexible manufacturing systems may be needed to cope with it.
(f) *Competition*   The Japanese appear to be moving from low cost, comparatively high-tech products to the production of multi-technology products. Competition is becoming more and more knowledge-based. To compete, increasing investment in R&D is required.
(g) *Technology*   Plastic motor components and composite materials are quoted as moves into higher technology which will be occurring in the 1990s. Components will be lighter, more durable and cheaper. Improvements in the use of computer-aided manufacturing, with CAD, CAM and Flexible Manufacturing, will be important.

The possibilities quoted above appear to be realistic, but with exceptions. Product proliferation has probably killed more companies than it has saved. To make a basic product and then paint it in different colours is one thing, but to make many different products is totally inadvisable and should not be attempted.

The premise that only companies making high added-value products will survive is again debateable. Flexibility and nimbleness are admirable, and complex products are also a positive gain—if they cannot be duplicated quickly and cheaply in Taiwan or South Korea. But what is a high added-value product—a radio tape player? These are made in Singapore far more cheaply than is possible in Slough. Technology can be duplicated easily anywhere in the world. Technology therefore will not save the West's manufacturing activity.

There is still room in the UK for well-run, semi-smokestack industries, with tough financial controls and solid, well-planned investment.

2.2.3 World automotive components product market developments
Motor components, their production, sourcing and supply, provide an excellent example of how a dynamic production market is developing and will continue to develop during the 1990s. What are the philosophies, the customers, the future sourcing activities which could come about?[†] These are some of them:

[†] 'World Automotive Components.' *Financial Times*, 8 June 1989.

(1) Competition intensified throughout the 1980s and will continue in even more brutal fashion in the 1990s.
(2) Sourcing by motor car assemblers is done on a world-wide basis. Rarely, if ever, is a local supplier preferred because he is a local supplier.
(3) Motor component suppliers and car assemblers have grown closer and closer together. Service and quality are now, probably, more important than price, though price, once negotiated, is very difficult to move upwards.
(4) Component suppliers are now making more and more modular assemblies rather than single components. Much sub-assembly originally carried out in vehicle-producing factories is now done by suppliers.
(5) Establishing a source of supply nearly always entails the right of the buying company to review manufacturing processes in the supplier's factories, to ensure that quality is correct.
(6) Much of the stock-handling, once done by motor assemblers, has now been forced back into component suppliers' factories. They in turn have had to constrain stocks within their own raw material suppliers.
(7) Motor component suppliers have tended to follow vehicle assemblers round the world. Because Toyota sets up a vehicle assembly plant in Derbyshire, there is no guarantee that Japanese motor components suppliers will not follow. They probably will. Globalization is a vital factor in the world car component industry.
(8) Competition forces every supplier to search diligently for cost reductions, better design, improved services, all the time.
(9) High technology is being forced on component suppliers so as to achieve strategic fit. Perhaps it can only be achieved in this way.
(10) Preferred supplier agreements are becoming essential in achieving any major piece of business. How to become a 'preferred supplier' should exercise the minds of component suppliers every minute of the day.
(11) Around 60% of all supplies are sent to original equipment manufacturers: the rest to the various after-markets. This proportion dictates the need to gain OE contracts above all else.
(12) For the UK motor component industry, 1992 arrived some time during the late eighties.
(13) The UK had an average pay rate some 70% below that of West Germany in the mid-1980s, but West Germany had productivity which made it just possible to be economic making components in that country. A comparatively small change in either productivity or pay rates could alter the manufacturing sourcing of components. This will continue to be a matter of calculation and decision-making.

2.2.4 The general response to changing market demands

2.2.4.1 Strategic fit
Some definitions are necessary at this juncture.

(a) Product market—the relationship between markets, market segments and the products designed to serve them.

Sec. 2.2] **Pressures to change** 31

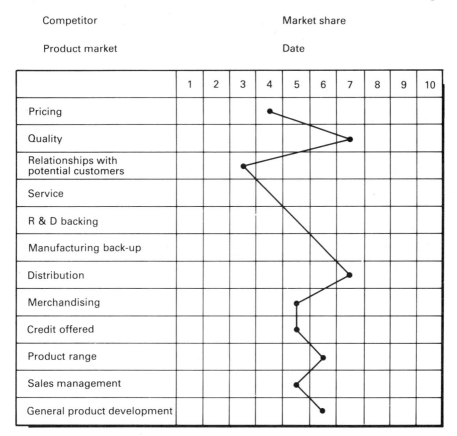

Fig. 2.1. Competitor competence profile.

(b) Strategic fit—the establishment of the relationship between markets and associated products, with all the aspects of manufacturing and distribution which should ensure that products are sold in planned quantities, in requisite markets and achieve appropriate profit margins.

(c) Competitiveness—having a strategic fit which is superior to that achieved by other organizations who wish to sell in the same markets and market segments.

Key analysis in establishing a strategic fit is to produce competitor profiles. These are graphical analyses which compare product market strengths and weaknesses of competitors with those one's own company (see Fig.2.1).

The product market system (Fig.2.2) shows an appropriate relationship which should achieve requisite sales. There are a variety of inputs. Some, like the product range, should follow from considered strategic decisions. The most important—cost: price, service, stocking levels—will be derived from production planning and manufacturing activities.

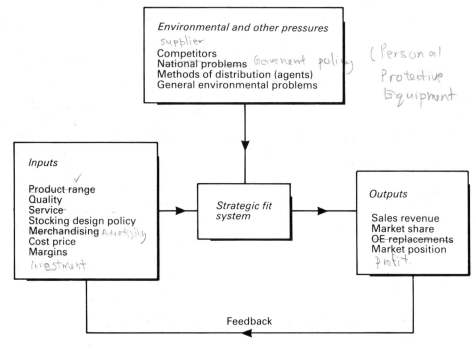

Fig. 2.2. The product market system.

2.2.4.2 Obtaining strategic fit
The relationship between the manufacturing process and the product market can be starkly expressed via establishing 'strategic fit'. An example will illustrate this point.

A motor components manufacturer in the UK would like to sell its products in Germany. What strategic fit is required?

The product market requirement

(a) A product range which covers 95% of the German car parc.
(b) Zero-defects on products sent. High specifications demanded.
(c) At least a 95% service level over the whole product range, slightly higher on fast-moving items.
(d) Price no higher than middle rate of German supplier, with the possibility of only modest (say 2%) price rises in each of the next five years.
(e) Guarantee of no interruption in deliveries, once these have commenced.
(f) Product performance characteristics to be superior to those currently offered by local German manufacturers. Precise details to be agreed with local users.
(g) Merchandising and packaging to be superior to that provided for the UK home market.
(h) At least 10 main distribution points in West Germany with the possibility of using up to 20 agents.

(i) For original equipment users, a detailed record of the tests carried out on all the products sold, plus allowing visits from quality specialists to ensure that total quality management is in place and is effective.
(j) Products to conform and pass local industry appraisal tests.

Possible responses

(a) Determine tooling requirements to provide a suitable product range. Calculate cost and how long it will take to have all the tools ready for production. Who will make the tools?
(b) Consider the impact of making the needed product range on:
 - Machine capacities
 - Machine utilization
 - Cost of machine running
     set-ups change-overs
     — materials.
(c) How can zero-defects be achieved? Introduce total quality management. How long will this take to become effective?
(d) Quality generally. Are the quality requirements of German customers the same as those in the UK? If not, how do they differ? At what intervals will German customers want to visit the factory? What will they want to see? What written procedures do they expect? Can these be provided? How soon will it be before an appropriate procedure is in place?
(e) Cost price. The forecast price increases suggest strongly that if contribution margins are to be sustained, direct production costs can only rise by an amount equal to the product price rises. While extending or improving contribution in total will obviously help, a constant reduction in contribution margins can be highly dangerous. Only year on year improvements in productivity will close the gap between forecast price rises and cost increases. Everyone should know this, especially shop stewards and operatives.
(f) New technology, perhaps robotics, will help to reduce rejects, work in progress and costs generally. What new technology is needed? How much will it cost? How is it to be funded? What usage factor is forecast?
(g) What changes in culture, in relationships, in communications, in pay and motivating factors, are needed to ensure that deliveries will not be interrupted?
(h) While stocks in the factory can be minimized, quite large quantities of stock will be needed to service the replacement market. How much? How will this be funded? How much will it cost to hold? How much will it all cost?
(i) Are planning and stock control systems good enough to ensure the service levels required, even if the factory is geared up to provide the flexibility envisaged?
(j) The degree of sharpness and flexibility required is more than UK companies require. How is this to be achieved? Can management initiative and skills be brought up to the levels needed? What new training is required?
(k) What extra controls must be put into place to ensure that strategic fit is achieved consistently?

The plan
Any product market forecast should be supported by the strategic fit requirements. This will be especially true where new markets are being entered or where competition is fierce and getting fiercer. The manufacturing framework should provide the basis for establishing a plan. Strategic fit requirements should be entered under the framework elements:

Technology
Resources
Work Organization
Systems
Payment and Motivation

The sales forecast should be verified via the strategic fit activity (see Fig 2.3). It has often happened that salesmen have made forecasts, which in reality have no possible chance of being achieved, because the necessary framework relationship has not been put in place.

Strategic fit requirements Plan 19				
Product market	Anticipated revenue contribution	Strategic fit requirements	Items not provided	Plan and timescale cost/benefit

Fig. 2.3.

Sec. 2.2]   **Pressures to change**   35

The aim of any plan should not be to gain revenue or even profit. It should be to sell installed capacity to the customer or customers who provide most contribution.

2.2.4.3 Strategic fit and the Single Market†

The aim of the Single European Market is to achieve the free movement of goods across all the old national frontiers of EC members.

No matter whether a company has exported to Europe before or not, the Single Market will have changed sales selling and competitiveness forever. Everyone is now in a European business. The corollary of assuming Europe to be a 'home market' must follow. Whether a company is more vulnerable to competition must be part of a debate within the company.

For companies in the motor components industry, exporting to Europe has been part of life for a long time. However, these companies must review what they are doing, taking the following into consideration.

Marketing Standards

Standards will be a key element in ensuring that goods can be bought and sold. Adopting the common European standards will make sure that organizations will be able to satisfy common regulations. For example, products must pass certain safety tests.

Directives setting out 'essential requirements' will apply to all products, irrespective of whether they leave their country of origin or not. The directives will apply quite severely on those products which will be used in public service. So, anyone anticipating submitting proposals for a public service contract will need to follow the appropriate directive precisely.

Quality standards

The directives will influence every aspect of product quality from manufacture to after-sales service. For the producer, this will mean a more efficient, perhaps more cost-effective manufacturing process. For the customer, reliance on product quality should play a large part in the purchasing decision.

'Quality' control also has an impact on product design. The assessment of company quality systems in the UK will normally be done under the guidelines and procedures set out in British Standard 5750 (which is identical to European Standard EN 2900 and International Standard ISO 9500).

Action plans on Standards

Manufacturers should know what Standards apply to the products they make and sell. Whether the Directives are actually promulgated or merely in a draft form does not really matter. Products and processes should conform, though it seems possible to influence the standards which are being formed.

Directions which have been adopted are published in the *Official Journal of the European Communities.*

---

† A useful set of pamphlets is published in the UK by the Department of Trade and Industry.

Manufacturing

The usual procedure will be to manufacture in accordance with promulgated Standards; though manufacturers can use other procedures, provided the end product complies with 'essential requirements'.

'Attestation' is the requirement of manufacturers to show how their products meet the 'essential requirement'. Independent test results will normally be required.

UK policy is that if a Directive permits a choice as to whether an independent body is needed, then the manufacturer can make his own mind up about test results.

Free sale

A product which conforms to the 'essential requirements' can carry a CE mark. This will ensure that a product so marked can be sold freely anywhere in the Community.

Transitional arrangements

If a relevant Directive has not been promulgated, a local national Standard can be used instead, if the EC has agreed.

Testing

Testing laboratories need formal accreditation and will be assessed by technical personnel, expert in the product procedures which will be tested.

Product certification

Several European countries, as well as the UK, have adopted product certification schemes. Formal assessment of production methods and quality assurance procedures are normally needed.

General marketing strategies

The Single Market provides major opportunities to increase sales, though some companies may be faced with far worse competition than they have suffered previously. Sales will only increase if the following factors are taken into consideration.

(1) Europe is now considered to be the 'Home Market' with all this means in marketing and selling.
(2) Unlike the UK home market, manufacturers should gain knowledge about trading conditions in each territory, including customs, any methods of selling intrinsic to the area and any product certification required.
(3) Methods of selling should be reviewed, including establishing agencies and partners.
(4) Management should be re-directed towards achieving increasing competitiveness and sales in the Single Market.
(5) Bidding for public service contracts may not be possible where previously it was.
(6) All competitors throughout the Single Market should now be known and competence profiles determined for them.
(7) Consider changing the product range to satisfy new customers.
(8) Sales will still depend on quality, cost, price and service.

Sec. 2.2] **Pressures to change** 37

(9) Foreign language documentation, sales literature and invoicing may be necessary.
(10) After-service could be a problem.
(11) Distribution becomes more complicated. Are new warehouses needed? What stocks should be carried?
(12) Can local organizations handle deliveries and sales?
(13) Product development may need to take product certification into account. Development may need to be speeded up.
(14) New technical standards may demand new production processes
(15) Relocation of plant may need to be discussed and developed
(16) Capacities may need to be changed.
(17) Language skills may need to be enhanced—rapidly—as well as the appreciation of national cultures.
(18) Identify the increased costs which will follow from trading in the Single Market:
    Transport
    Distribution
    Certification procedures
    Later payment
    Extra stocks, working capital and financing
    Agency fees including advertising
    Sales promotion
    Market research
    Training.
Taking these factors into account, will unit costs be appropriate to increase sales?
(19) Currency management—a new idea? 'Export management' will probably be needed though it will be a 'home market'.
(20) Information systems may be needed which enhance marketing data currently achieved. IT may be crucial.

2.2.4.4 Niche marketing Focused business Focused factories

The second of the academic's proposals (section 2.2.1) to ensure success in product markets was 'niche marketing'. This seems just an extension of the traditional market segmentation which most companies do.

Strategic fit needs to be carried through to its logical conclusion by relating market segments or product markets to:
  businesses
– factories
– product lines
– systems administration.

Setting up focused factories, business etc., should provide these benefits:

(a) The business becomes completely profit-oriented. No one should act as simply a salesman, an administrator, or a production manager any more.
(b) The 'corner shop' mentality is produced, with substantial improvements in culture. Saying that businesses are to live or die by their own results produces noticeable benefits in performance.

(c) Resource allocation is improved. Capital investment and working capital would need to be bid for and given to the bidder who promised or achieved the highest return. (This may not be true at all times.)
(d) Diversions into side alleys of a non–profitable nature should cease.

2.2.4.5 *Globalization*
It all makes a lot more sense if globalization is included. Having a global spread of production marketing facilities with local cost selling advantages considerably improves the chance of tackling local markets effectively.

Globalization as attempted by many British motor component suppliers has been more than matched by the Japanese, who have gained considerable benefits for their car industry. These have been:

(a) A direct reduction in costs has resulted due to extended production.
(b) Rationalization of product runs though some variation is allowable
(c) Possibility of shorter product cycles as amortization of tools has improved owing to increased use.

The Japanese have scored because they can afford to make model changes far faster than their competitors in the USA and Western Europe.

Other possible effects from taking globalization as a starting point for product markets are:

- *Cooperation* This may be possible with local suppliers/manufacturers/agents, without the need to establish one's own unit in the country concerned.
- *Rationalization* The 'world car' has been a phrase long used by vehicle assemblers and component suppliers. Future rationalization is likely to take place to make one vehicle design suit a variety of countries and conditions.
- *EC 1992* A European market will become a 'home market' for every one in Western Europe. Competition will increase but so will opportunities.
- *Organizational change* Organisations need to change to ensure that due regard is paid to local cost/marketing advantages especially in product marketing. The problems of what to leave as a centralized service and what to decentralize to obtain strategic fit could be quite complex.
- *Investment* Investment in specific markets should be made using national rather than international opportunities. Operation on international funding basis often causes a great deal of pain. Finance directors are not always brilliant at organizing international loans.
- *Size* Size will become increasingly important in tackling global markets, but this problem might be overcome by buying technological production marketing agreements with friends, rather than own arrangements.

However, globalization can cut many ways. South Korea, Malaya and Brazil all make reasonable motor vehicles; components should soon follow. Competition from European manufacturers is probably the least of a UK supplier's troubles.

2.2.4.6 The home market

Having declared that globalization is important in considering a product market strategy, it is perhaps paradoxical to say that unless a supplier can dominate his home market, he is unlikely to become significant outside it.

The home-based supplier has all the advantages—language, geography, local culture, nationalistic preference—on his side. Failure to capitalize on these factors could be disastrous.

Beral in 1985 exported to countries that Ferodo export managers could scarcely find on the map, yet their presence in the local West Germany replacement market was too small. They had to sell in their home market to stay alive. They did not, and paid the penalty.

The British car assemblers started to go downhill fastest when they thought they were practically inviolable in the UK. In 1960 when import penetration was negligible, they were at their most complacent. They then lost the battle with importers despite all their apparent geographical, cost and nationalistic advantages.

Exporting needs a hard, often long and frustrating, period before it becomes profitable. Home markets often have higher margins and easier relationships with potential allies. Home markets are not dependent on fluctuating exchange rates. Unless local suppliers can, therefore, make use of all their potential advantages, they will not win elsewhere. *A manufacturer must dominate his own local market.*

2.2.5 Market share

It has generally been accepted that the seller with the biggest market share has considerable advantages over those with less. Hopefully, the market leader is able to dominate product pricing, maximizing his contribution. His marketing activities will tend to be followed. As soon as a newcomer attacks the market, the leader's dominant position will allow him to take action which will nullify the attack—or so the textbooks say.

If this were really true then the Rover Group would still be the premier motor assembler in the UK. Any number of British companies would still be in existence and sitting on a dominant market position—everyone from Alfred Herbert in machine tools to Coventry Climax in fork-lift trucks.

Of course it is not true. Japanese companies like Kumatsu have been able to consider carefully what share of the market they want and then set plans and carry through actions to achieve it, waiting patiently until success is won. Patience is the key, once strategic fit has been achieved.

There is no doubt that a high market share can bring advantages. Japanese companies have proved conclusively that it is possible to move from a position of negligible proportions in a market to one of dominance. Many British companies have proved the opposite—market dominance can very quickly be undermined.

It is probably curious, then, that Ferodo rarely set out, as such, to achieve $X\%$ share of a market. However, every major OE supply contract that came up was fought for bitterly. It was not that a 100% market share was sought—just that it was known that some contracts would necessarily be lost. Perhaps this, in an OE situation, is the right strategy to follow.

Market share can be won by being a better competitor. If this largely means reducing price, profitability obviously will fall, unless total unit sales increase. This is why calculations of contribution are so important. It is not price so much as contribution earned that matters. While this continues to increase, then all might be well.

The problem with trying to advance market share is that it may cost profit in the short term. In a company that demands immediate profit, the time-scale for improving market share may be too long, despite the potential benefits.

So, people in charge of businesses, product markets and marketing should always know that they are employed to maximize contribution both in the short and longer term. This should be a useful indicator as to whether they need to chase market share more vigorously.

2.2.6 Competitiveness

A recurring theme among our salesmen in the eighties was that we had to become competitive. How precisely did they define competitiveness? It appeared to relate to the following:

- Have better products than competitors.
- Have better deliveries than competitors.
- Achieve a 100% stock service.
- Extend the product range to include anything a customer or potential customer had ever asked for.
- Achieve one business's products for sales at the expense of other businesses.
- Ensure the factory had infinite flexibility, including the ability to make one item at a time on the day it was requested.
- Maximize the advertising budget to support selling campaigns even if this meant reducing costs in, say, production engineering.
- Extend the marketing department and pay its people more.
- Ensure that no poor-quality material ever left the factory (i.e. achieve zero-defects).
- Sell at prices lower than competitors would match (or want to match).
- Ensure that appropriate contribution was earned, by making the factory reduce its costs on a year-by-year basis.

No doubt these were genuine, even heart-felt requirements, from people who were under considerable pressure to sell more. Enough will be said later about quality, and later still about MRP II's ability to improve service. Perhaps what remains is to debate manufacturing costs and the production and service achieved.

Over the years, production managers have grown weary of hearing salesmen say, 'Our costs are too high.' Yet the way marketing selling occurs can increase cost considerably. Making the best use of installed capacity should be as much a sales marketing problem as a production one. Salesmen are in business to sell installed capacity to the highest contribution bidder. This should be a key objective for them.

Being competitive is the achievement of a strategic fit with product markets which will maximize contribution from installed capacity. It is not about being the best in the world, or about having levels of service, quality, price, etc. which competitors are

Sec. 2.2] Pressures to change 41

not able to achieve. There are still trade-offs between costs and quality, service and product range, which sales marketing personnel need to consider carefully. Without such trade-offs being calculated in a product market context, things can go badly wrong.

To end this review, the production manager might reply to the salesmen that his own competitive ability will be enhanced by:

- achieving the sales forecast in the product item and potential margin put in the profit plan;
- filling installed capacity;
  obtaining orders or delivery on a timescale which does not force overtime;
  obtaining orders which minimize set-ups;
- a range which maximizes productivity.

2.2.7 High added value in product market strategy
Following an interview with Colin Hope, Chairman of T&N plc, the *Professional Engineering Magazine* of June 1989 gave a brief review of what T&N's marketing strategy had been since Lord Tombs had taken over in 1982. It was as follows.

(1) T&N planned a move away from low added-value products like building materials, into higher technology products, where materials science was important.
(2) AE (an acquisition of 1986) did not at first fit the concept of small well-run units making niche-type products. It was too centralized.
(3) Products are now specialized (as far as possible), e.g. Hepworth and Grandage's petrol and high speed diesel pistons.
(4) T&N attempts to be big enough in each field or niche to be in the world league—a concomitant of survival in the 1990s.
(5) T&N's belief is one which combines advanced materials technology with component design.
(6) Sales are generally made to component assemblers (e.g. brake assemblers) and agents and distributors, so avoiding much sales/distribution/administration costs.

2.2.8 Competitor competence profiles
We used the traditional competitor competence profiles to show how competitive we thought we were. Fig. 2.1 shows a typical example.

When a competitor competence profile is carried out by sales personnel, there is a tendency to overrate competitors and underrate company achievements. They often use the occasion to air their grievances about company failings. Achieving an objective appraisal is usually difficult.

Company profiles can be overlaid against those of competitors, and differences determined and discussed. While useful in setting out discrepancies in competence, profiling does not (and probably should not) indicate the relative importance of the various elements. Merchandising may be of limited importance, for example, and the fact that a competitor has eight points for it may not necessarily be relevant, whereas pricing could be vital in gaining market share, and the company may be doing well in this respect.

2.2.9 The product range
A broad strategy for a product range in Ferodo was established as: To supply, but not necessarily to stock, a product range based on cars and commercial vehicles assembled in Western Europe and Japan in the last ten years.

Data was collected under the following headings:

Vehicle manufacture
Product group
Vehicle model
Other associated models
The catalogue reference accepted by the trade
Local product group number
Material type
Main subsidiary markets
Vehicle sales
Predicted vehicle parc in the next three years
Current market share
Likely demand forecast for next three years
Recommendation whether the products should be
    made
    bought out
    stocked
    sold

If stocked, what category should the item be (therefore what service should be attempted).

2.2.10 Forecasting
Among all the causes for the antagonisms which perhaps are inevitable between sales and production personnel, the failure to produce a forecast which comes anywhere near reality must rate very highly.

Forecasting is difficult because:

(a) Sales people, like everyone else, prefer objectives which underestimate their potential. Caution is all very well, but when it means that conflict over available capacity occurs half-way through the year, it is counterproductive.
(b) To many production managers steeped in MRP II, management accounting, team working, added value, etc, salesmen appear to be the last in the organization to become 'professional' The chairman in the sixties once said to me, 'We don't want techniques people like you in the company, we want people with flair.' He may have been referring to a typical sales manager, speaking no foreign languages, returning to the company from Germany with a $2\frac{1}{2}$% price rise but hoping for a 7% pay rise. Flair, even among a sales force, is never enough.
(c) It is difficult, but not impossible to, say, extrapolate car parc data into the real world of agents and key customers.

(d) Even if product markets can be forecast with reasonable accuracy, doing the same for individual products can be practically impossible. Yet unless individual items are forecast, a calculation of work content again is impossible.

What then is an appropriate way of establishing this most important element in looking at product markets?

As part of introducing MRP II, we wrote a user specification. This laid down the information users were likely to need in the next decade. It established relationships in information between functions. It produced a means for establishing a database from which MRP II and all its related systems could be operated.

Forecasting then received a rigorous examination, and the following requirements were listed in the User Specification:

General
  Volume changes from previous years by product markets
  Price changes from previous years
  Resulting capacity changes
  Changes in material types and associated volumes
  Changes in capacities

Product market
  Likely demand arising from designated:
    product markets
    — customers
    — agents

Product groups
  The same as previously

Data needed
  Likely revenue for the year
  Likely revenue for each month adjusted for seasonality
  Prices per year or per month
  Average standard price per piece
  Contribution per product group market
  Percentage revenue contribution
  Revenue contribution for agreed major customers.

Forecasts should be rolling, updated each quarter for a year ahead. In the short term they should include a statistical forecasting technique (e.g. exponential smoothing); but the longer-term forecast should be based on marketing intelligence, achieved contracts (OEMS), established off-take of key customers, desired market shares etc. (resulting from the application of top-down planning).

As products grow increasingly sophisticated, especially in the finishing operations required, it will be necessary to forecast individual items. This can best be done by using car parc data.

Inputs to the system
In establishing a satisfactory forecast, considerable attention should be paid to:

- The external environment, i.e.:

    gross national product forecasts
    public authority spending in specific product markets
    stocks
    exports
    imports
    exchange rates and changes to them
    wages and earnings
    prices inflation
    wholesale retail and consumer spending trends
    raw materials

- Economic trends in major industries; such as:

    transport
    motor vehicles
    iron and steel
    farming
    banks and lending
    building
    textiles

- Product market changes:

    characteristics
    capacities & technologies
    size
    profitability
    strategies—used and probable degree of success
    share of market and desired share
    service needed
    key customers in each product market

- Product life cycles

    position
    growth or decline

- Product contribution by:

    market segment
    product family and product
    salesman
    export market
    home market
    key customers

- Technical changes in:

    products
    markets
    end users
    materials used
    distribution techniques
    data processing

- Potential impact on:

    quality
    profit
    function
    end users
    product design
    market competitors

- Market competitors in product market. Their:

    number
    market share
    competence in

        technology
        quality
        service
        merchandizing
        pricing
        distribution.

What is being done to wrest market share from competition?

Strategic fit
- Competence related to product requirements
- Graded discrepancies in strategic fit
- Possibilities/action plans to create strategic fit
- Works environment and revenue earning:
  automation/technology
  capacities
  skilled/unskilled employees
  Health and Safety requirements
  new forms of work organization
  (all leading (or not) to service/quality/cost improvements)

- Export potential:

    country-by-country analysis of
        economies

culture
markets
competitors
technologies

- Own expertise in:

marketing
   publicity
   pricing
   product design
   market research
   distribution

production

   productivity
   machine design
   material usage and control
   general resource deployment
   materials handling
   engineering.

Processing
The input data outlined above should be input into the computer system, the result being a seasonalized forecast, reliable within quite narrow limits—something that production personnel occasionally dream about.

Interfaces
Data needed for the forecast and the result should interface with many other product market and profit planning routines, notably the following:

- The product market plan:

    market share in car parcs
    key markets
    contribution needed
    revenue needed

- Customers orders: demand by product group

- Sales analysis/ledger:

    revenue per key customer
    contribution per key customer
    business potential
    business received.

Output
(a) The main output will be a yearly forecast for each product group, showing by quarterly value and standard each market's future demand.
(b) It should also be possible to publish a price index on a year-by-year basis, showing changes from a base year for each market.
(c) In the same way a volume index can be produced.
(d) Contribution per product market would also be output.

2.2.11 Marketing Strategic Fit Activities

The preceding parts of this section have suggested that marketing is not an arcane subject understood only by well-trained specialists. The product market strategic fit debate should be one in which all managers in an organization should take part, because they all need to be involved. Production managers should not be faced with a *fait accompli*, where they are asked to perform minor miracles, with fewer resources every day, because the marketing people say this is what is needed.

Rationality can be ensured by using the simple form shown as Fig. 2.4. Once an objective is set, column 1 should show every detail needed under the product market headings e.g.:

product range
packaging
service
quality
selling structure
advertising

Objective: To obtain 20% of West German replacement market		
Product market	Strategic fit	Actions/timescales
Product: *Car disc brake/pad* Product range of German vehicles	Not available. Inadequate tooling	To be built up: • tooling • stock
Quality KBA approach	Non-existent	To be obtained
Market: *West Germany* Service 97% of total range always available	Possible once range is in place. Use of Cascade system	No further action needed if stock control system is utilized as a product market requirement

Fig. 2.4. Strategic fit statement.

The list might stretch for two or three pages and go into considerable detail.

Column 2 should then list the current strategic fit. How far, for example, does the current product range match that required? If it does not in some way, then the actions needed to ensure fit are quoted in column 3.

The key elements in this activity are:

(a) making the best possible description of the product market requirements, if possible putting numbers, rather than generalization;
(b) debating the product market data with all managers and justifying the requirements;
(c) ensuring the strategic fit is set down as a direct cross-over from the product market;
(d) allocating responsibility for the actions, putting a cost and a completion date on them.

In reality, a further analysis is then needed in which the actions are listed, the people responsible named, the completion dates agreed and then the resource cost to be used is compared with the quantified benefit shown, all done as carefully as possible.

2.2.12 Customer relationship—quality assurance

For most motor component suppliers. during the last few years the major change in requirements between the suppliers of the components and their users, has been in the drive for higher quality. Five or six years ago it was probably unheard of for a motor components factory to be visited by a car-assembler's staff, to discover whether there were adequate quality assurance procedures in place. Now it is nearly an everyday occurrence.

Without a suitable grading by the user, the potential supplier is left out in the cold. Suppliers ignore aspects of quality assurance at their peril. Quality assurance is therefore a vital component in product market planning and that is why it is included in this section.

At the Lucas Automotive supplier conference in 1988, the quality requirements were pronounced as:
- quality standards
- failure modes and effects analysis (FMEA)
- capability studies
- process control plans
- statistical evidence of control
- annual re-validation of components
- certification of conformance
- packaging and labelling specification.

If any of these requirements were not met, it was said, products would be rejected.

Lucas suggest that a total quality partnership is needed between themselves and their suppliers, with:

- common aims and aspirations
- mutual trust and cooperation

- desire to improve continually the product and services
- clearly understood responsibilities.

The policy of the supplier must be to have 'zero-defects.' To achieve this position, it is likely that the supplier's system included:

- A process where continual improvement is regarded as the norm
- Progress made through advanced technology and, above all, people,
- Quality systems based on self assurance, i.e.:

    - quality needs to be built into the manufacturing process
    - the operative is 100% responsible for quality
    - the next operation is considered to be a customer
    - SPC,[†] must be in place and effective.

- Teamwork between quality and manufacturing people at all stages of the process. Quality is everybody's job. The days of having 100% inspection are over.

    Lucas automotive SQA department has gained authority to carry out the following functions with its suppliers:

- appraise and classify suppliers
- maintain a suppliers' classification register
- maintain a suppliers' performances rating
- provide supplier improvement assistance
- assess quality planning activity.

The obvious aim of all suppliers must be to go on the 'preferred' list, which necessitates a greater then 94% overall rating. Bidding to be a Lucas supplier therefore dictates that the supplier should use, effectively, advanced quality planning techniques. The suppliers need to have similar quality concepts to Lucas and its customers, so that a long-term relationship can be established or maintained.

Ford have a formal programme of 'Advanced Quality Planning' which in essence catalogues the key features of a new product manufacturing development cycle. It covers:

Feasibility analysis — suppliers should show that they are capable of manufacturing, assembling, testing, packaging and shipping to acceptable levels.
Specification/design FMEA
Proprietary items
Processes
Controls
Tooling
Gauging
Machine capability studies/SPC/process potential

[†] SPC is statistical process control.

Engineering specification testing
Sample submission
Packaging
Sub-contracting
First production shipment

Ford carry out an SQA systems survey and write potential suppliers a report where a rating is given from each review element. The main elements are:

- Responsibilities for quality planning
- Product auditing responsibilities
- Written procedures
- Implementation of written procedures
- Written inspection and lab test instructions
- Gauges
- Controls used during product processing
- Plan and FMEAs
- Statistical methods
    SPC
    machine capability studies
    – out-of-control conditions

- Incoming materials

    – evidence of SPC in suppliers activities
    systems used by supplier

- In process and outgoing procedures
    inspections
    – corrective actions

- Repair and scrap procedure and analysis of non-conforming products
- In handling storage and packaging.

To the Lucas and Ford requirements can be added these:

(a) There is a need to identify and meet the quality assurance system requirements of all customers including those requiring compliance with relevant sections of BS 5750, ISO 9000, EN 29000 or AQAP series of quality systems Standards.
(b) EEC directions on product liability, embodied in the consumer Protection Bill (1987), need to be taken into account.
(c) Motor manufacturing demands for total quality management systems must be considered.

All this should lead to the production of a quality policy manual and company-wide training in statistical process control. The aim is to achieve company-wide changes in attitudes towards quality. The policy document should cover:

(a) *Company objectives*
    Profit and quality go hand in hand

Quality must be an intrinsic part of:
    product development
    raw material and component supply
    production
    customer services
(b) *Quality definition*
    System must ensure prevention, not cure
    Performance standard is zero-defects
    Measurement is by cost of quality
(c) *Company commitments*
    Products sold will be only those made to agreed specifications
    Quality is company-wide
  Senior management must constantly reinforce quality needs
    Right first time
    Team work
(d) *The quality system*
    Purpose
    Definition
    Implementation
    Systems elements, e.g.:

- contract review
- AQ Planning
- design management inproduct development
- control over internal and external factors

    Demonstrate conformance
    Preserve product quality until arrival at customer's premises
    Customer support—after sales
    Audits
(e) *Responsibilities for*:
    Sales and marketing
    Product development
    Administration
    Production
    Q.A.
    Research
    Engineering
    Personnel
    Finance
  need to be determined and related to one another
(f) Quality manual
(g) Policy Review Team

For a company which had prided itself on being a quality supplier, the failure to gain the best possible quality rating from, say, Ford or Lucas, came as a considerable shock to us. It had not been seen as a great tragedy to throw away 5%, 10% or even

15% of a batch, so long as the remainder was perfect. But of course it never was, and there was no guarantee in using even 100% inspection that it ever would be.

Of all the failings in our organization, this was perhaps the most classic. Rows of operatives, acting like Horatius on the bridge, carried out final inspections. Operatives hurried to complete orders on time, perhaps giving only a passing glance at quality. Some patrol viewers wandered harmlessly around, making minimal impact on quality.

These activities could never improve quality. Costs of:

Time spent sorting batches of material
Rectification
Rejects
Material losses and poor utilization
Labour, tools, energy were rising inexorably.

The mistake of believing that detection rather than prevention of rejects was important, must have cost millions of pounds.

2.2.12.1 *Statistical process control*

Statistical process control is a useful peg on which to hang quality assurance. In itself, the statistical process is nothing new. It was being well applied in Metal Box Co. in the mid-fifties. It is the surrounding activities which make it important.

(a) Process control needs:
 – a clear quality specification;
 – checks on machines, processes and individuals to ensure that products will be made to specification;
 – audit equipment to ensure the right quality is being achieved;
 – support and training of supervisors and operatives to ensure that high quality products are made.

(b) Statistical process control (SPC) is an activity which measures both process and products, to ensure conformity to specification. Statistical information should establish the actual and trend deviation from standard. The outcome should be to produce items within defined limits.

(c) Key objectives will therefore be:

 – to reduce variability in processed products;
 – to increase consistency, predictability and dependability.

(SPC should prevent defects, as opposed to inspection which detects them.)

(d) General statements pertaining to SPC:

 – The process is a combination of people, material and processes.
 – Action is taken on the process, not on output.
 – The aim of SPC is never-ending improvement. The best quality rather than a minimum level is the aim.
 – Products must be made 'right first time' with 'zero-defects'.

Sec. 2.2] **Pressures to change** 53

Special disturbances in the process (e.g. tool failure) have to be eliminated before SPC can work. Common disturbances are inherent in any process and can be handled by SPC, e.g. machine properties, humidity, etc.

(e) In the statistical process both variables and attributes are measured.

Variables include length, thickness, diameter, density.
Attributes include blemishes, discolourations and other aspects which cannot be measured accurately.

Control charts
The variable data is plotted on a control chart, at predetermined intervals—possibly every 15 minutes or one hour, depending upon the potential variability. If charts are completed by the operative, he will see when the process is going out of control and can immediately make necessary adjustments to bring control back.
Consequently, the chart will help the process to perform in a consistent manner, with lower variability and will indicate the difference between special and common disturbances.

Data
Normally this will be 'mean' and 'range', i.e. the arithmetical mean and the range in terms of distribution of results.

2.2.12.2 *Gaining quality awareness*
Creating a situation where everyone is quality conscious is exceedingly difficult. Years of payment by results, based on output only, is a very strong conditioning mechanism to prevent the effective use of QA.

We introduced payments for good products only, in one part of the factory, and were then surprised that the reject level did not go down. The operatives were quite frank. They still earned more by working in a way that produced very high output though having some rejects debited against their pay, than working more slowly and having no rejects.

Like education in financial awareness, quality assurance needs a company educational scheme, followed by constant reminders of the need for quality, the effective and widespread use of SPC, and, in our case, the perpetual nagging of supporting evidence as follows:

(a) Product-line newsletters—these were produced monthly and were initially designed to support verbal communication about product-line performance of which quality, rejects and material utilization were a key part.
(b) Quality standard room—a room was set apart and a display of information on SPC, quality monitoring and measuring, cost of rejects, samples of good and bad work, information on customer complaints with examples was installed. Some of the display material was semi-permanent, some of short-term duration.

Initially, all operatives were given a period in this room with talks by the local quality engineer and production manager, with the long-term aim of encouraging operatives to go in of their own free will to find out what was the latest on the quality front.
(c) Weekly reject level display board.
(d) Quality poster.
(e) Machine identification—each machine where SPC was used was clearly identified and appropriate charts and records highlighted, using good quality plastic sheeting.
(f) General housekeeping—no factory which is bad at housekeeping can be good at quality; the two seem to go together. Improved housekeeping was an essential precursor to ensuring that quality was improved. The psychological benefits of the one were passed on to the other.
(g) Suggestion schemes—prizes were given to the best ideas to improve quality.

Good though these initiatives appeared to be, the best methods of projecting quality awareness were the SPC data established at each machine and a 'reject quality table' established on the production line. This table was used to record current reject rates and samples of poor quality products picked up in the production process. Operatives could see that the results of their poor performance were put on display when, hopefully, peer group pressure would ensure that an improvement took place.

2.2.12.3 *Quality circles*
'We are extending our quality circle programme not only for the obvious benefits it brings, but also because of the increased employee motivation levels on all aspects of quality that it produces'—so said John Egan, chairman of Jaguar, in October 1985, when talking about how quality was improved in Jaguar. At the same time, Ferodo was introducing quality circles and their activities were given wide publicity, especially in the various news letters that were then issued.

The quality circles failed to achieve anything of long-term significance. Why?

(a) The contention made in this section is that a major improvement in quality will only come about by company-wide changes in the perception of how quality should be assured and the variety of activities which management need to introduce. In the process, quality circles are only of marginal importance. They can help in generating interest awareness, but they are no substitute for management action.
(b) Quality circles, and the personnel who crew them, usually have considerable knowledge of their own immediate problems and what needs to be done to correct them. These, in the total company scene, may not be important.
(c) If the local shop-stewards see quality circles as a means of supporting management with no gain to operatives, they will not succeed.
(d) If first-line supervision believe that quality circles undermine their authority in running their unit effectively, then again success would be limited.

Quality circles might help generate a culture of cooperation and joint problem-solving, but it would be better if management were already instituting culture changes.

2.2.12.4 *The Deming philosophy and quality*[†]

The advocacy of one technique or the application of the activity which has apparently done well in Japan is highly dangerous. The training consultants are overburdened with wonder cures—just-in-time, total quality management, quality circles. They could do more harm than good, if applied outside a particular culture or philosophy. William Deming, an American, is an advocate of a 'philosophical approach' which the Japanese took to heart in the 1950s. The results are well known.

Deming put forward obligations which management had to follow if they were to be successful. Many, but not all, are recorded somewhere in this book, but their combination and presentation might be useful. The overriding view is that gaining quality is the key to success: if products are made which are 'right-first-time' then, automatically, productivity improves, costs go down, and customer service benefits enormously. What are the main ideas put forward in this philosophy?

(1) Consistency of purpose—the company should search for continual improvement. Yet, there must be stability within the company—once rejects have been eliminated and everyone from order clerks to production workers, becomes quality aware. Morale improves, relationships strengthen, insecurity of managers and employees should be eliminated.
(2) Emphasis on short-term gains debilitates quality and productivity. A longer-term view of people and quality is needed than the scope normally given by concentrating on monthly results.
(3) Mass inspection should be eliminated. Everyone should be an inspector.
(4) The majority of quality defects are not caused directly by employees, but by poor systems of control, especially statistical process control and how this is employed. Management's main job is to improve the quality system.
(5) The object of any business is to service customers. Their expectations should be exceeded rather than met. No company should just 'meet specification' or 'match competitors'. More than that is needed.
(6) Rarely, if ever, award a supplier a contract on the basis of lowest cost, alone. Reduce the number of suppliers. Control, if possible, their quality, and ensure that they have an approach to quality which at least equals that in one's own company.
(7) Training must be constant and to all levels of the company. Training may not be in a particular activity related to what an employee is currently doing. Self-improvement should dominate 'training' wherever possible.
(8) Leadership is a vital factor, instituting any change which is needed to improve quality.

[†] More information can be obtained from: The British Deming Association, 2 Castle Street, Salisbury, Wilts SP1 1BB.

(9) Emphasis on performance achievement, target setting, merit rating, should be scaled down, because, it is said, team activities are destroyed by it.
(10) Everything should be done to ensure that managers stay in the same company. Shifting around is not good for the long-term continuity which most companies require. Part of this process should result from much-improved communications and from the elimination of fear of the unknown which poor communication engenders.
(11) Eliminate targets and goals; substitute SPC which will show when a process is going out of control.
(12) Allow pride of workmanship to dominate the production process.
(13) Reduce or eliminate all traditional organization barriers which hinder the achievement of high quality.
(14) Eliminate exhortation to improve without giving operatives the means whereby improved quality can be gained.

There would be very few managers in the West who would disagree with the need to put quality high on the agenda of any management improvement programme. Would it lead, nearly automatically, to improvements everywhere? The proponents of 'just-in-time' think that exactly the same will happen from the application of minimum stocks.

The reduction in, or perhaps the total elimination of, targets or objectives is probably the hardest thing to swallow. These certainly have proved indispensable in the past. Even teams need to know what goals they are striving to gain, otherwise how can resource allocation be planned? Because the Deming philosophy has worked well in Japan, there is no reason to believe that it will fit absolutely in the west. To believe that it will is to fly in the face of the fact that transplanting techniques, even philosophies, from one company to another, let alone one country to another, rarely works out as intended.

2.2.12.5 Quality—the Juran philosophy

Dr Joseph Juran shares with William Deming much of the credit for improving the quality of Japanese goods. Juran defines quality[†] as those features of the product which respond to customer needs, plus freedom from trouble (i.e., re-work). The first factor is basic to selling the product; the second is fundamental to easing production problems.

The West, say Japan, has two problems in improving quality:

(1) It puts up with too much rectification and re-working.
(2) It has a wrong organization structure—i.e., major inspection departments.

Add to these factors, that senior management is not involved enough with quality, and poor quality perpetuates itself. Taylor-type incentives only add to the problem.

Managing for quality demands that each line department and local person is responsible for quality. Each department or line should have quality as a key objective.

† As in 'The Juran Method—don't delegate, do it!' *Works Management*, January 1989.

Teams should be set up to achieve them. Training in quality for everyone in the organization is essential.

Juran then recommends a 'Quality trilogy' consisting of planning, control and improvement.

- *Planning* includes identifying customers and relating their need to quality, if necessary developing different or better products to meet customer requirements, along with production processes that achieve good quality, under strenuous operating conditions.
- *Control* covers the choice of control subjects and the measurements of actual performance.
- *Improvement* encompasses proving the need for improvement, organizing the diagnosis of quality failure and providing remedies.

Each week in Ferodo, the site director helped in the diagnosis and improvements, by running a quality improvement team.

2.2.12.6 *Taguchi*

'We will do a Taguchi!' became something of a catch phrase in the company. Dr Genichi Taguchi (obviously Japanese) designed a method of quality manufacturing which he hoped would 'measure quality in monetary terms' and would show 'how to improve the quality without adding to cost. 'Quality is defined as any 'loss a product causes to society after being shipped, other than any losses caused by its intrinsic function'. Such losses largely reflect warranty claims, customers' dissatisfaction, or any side-effect which the defective product might engender. Loss can be measured by using a 'loss function', a Taguchi-derived, mathematical formula.

The Taguchi process is to make off-line quality engineering take the place of on-line quality assurance activity, with the main aim of reducing product variability.

The 'Taguchi' covers three elements:

- How the product is to be made.
- How the process can be improved to eliminate instability.
- How far, once the improvements have been determined, tolerances in raw materials on some aspects of the production process can be tightened up, so making the product more resilient to 'noise'.

It is in the second stage that Dr Taguchi, and our own engineers, believed that most benefit could come.

'Noise' is largely the reason why quality defects occur—either in the way the product is used, process noise, such as tool wear, and between product noise in imperfection in the production process.

2.2.12.7 *Relationships with Japanese organizations*

Speaking at the Motor Industry Conference[†] Mr M. Gower spoke on what the Rover group performance had been—poor quality, poor reliability and poor delivery. These,

---

[†] 'The Motor Industry in Britain. Which way in the Nineties.?' 19 November 1987.

he avowed, had been put right, and one of the most salutary experiences in the process had been the relationship with Honda in:

   product development
   manufacturing
   quality and reliability
   customer satisfaction.

Our own relationship with Japanese companies certainly forced a re-appraisal of what was satisfactory product quality. It is likely that the whole of our views on quality and on what can be achieved has gone up several notches as a result. This, for supplier, customer and end user, can only be a good thing.

CONTENTION

(1) The front line with the customer now starts in the supplier's factory. Any company that believes that sales or marketing personnel are the only contacts needed with customers is heading for trouble.
(2) It is now likely that the way the factory carries out quality assurance is as important in gaining and keeping new business as anything the sales force can do.
(3) The product market therefore is everybody's concern.
(4) It is likely that sales personnel are less professional in their approach than most other functional personnel in the company. The reliance on flair and intuition can be dangerous.
(5) Establishing a strategic fit between detailed market segmentation and company products and operating practices, backed up by competitor analysis, is likely to gain as much competitive edge as more sophisticated approaches.
(6) Going for market share increases may not be the best way to achieve high contribution earnings. Anyone wishing to initiate. a tough, comparatively short-term marketing discipline should institute as the objective: 'We are in business to sell installed capacity to the highest contribution bidder.'
(7) Globalization, for many companies, will be vital in the 1990s. Nobody should think that the home market is all that matters.
(8) However, if a company cannot dominate its home market, it is very unlikely to make a deep impression on its export potential.
(9) 'We must be competitive' could also be a dangerous cry from sales people. What is competitiveness? What must be done to beat competitors is of prime importance, not what is thought necessary to service every last whim of the customer.
(10) Quality and durability could be as vital as price and delivery performance in achieving sales.

A suitable set of strategies

(1) We will continue to enhance the use of focused factories, focused business and market segmentation to ensure that the strategic fit between company production and general business activities is as close as possible with customers.

(2) We will also continue to use competitor profiles and strategic fit analysis to make sure that the relationship between customer requirements and product activities is as balanced as possible.
(3) While competitiveness is largely concerned with quality, price, service and product range availability, reducing costs should be as much a responsibility of the sales force as that of production. Stability of demand and full utilization of installed capacity will go a long way towards reducing unit costs.
(4) It is essential that we dominate the home market. We must be able to outbid low cost imports, by price, service and quality.
(5) We will not chase increased market share at the expense of profitability except in the short term. Market share in the longer term must equate with enhanced contribution.
(6) We will have a product range which matches requirements of cars and commercial vehicles built in Japan and Western Europe in the last ten years. A product range database, indicating potential market shares, make/buy decisions and stocking requirements, should be built around this strategy.
(7) We recognize that globalization will be a major requirement within the industry and we are prepared to share techniques, production and marketing activities with potential allies, to secure an increasing foothold in global markets.
(8) We also recognize the need to concentrate on high added-value products. In our case this will mean using materials which produce high quality and longevity for which we can gain a higher than average price.
(9) We will ensure that quality assurance is accepted, understood and put into practice throughout the organization.

3

Technology

3.1 INTRODUCTION

A senior manager from a merchant bank visited the factory at Chapel-en-le-Frith in the late 1980s. He was shown the robotic cells manufacturing car disc brake pads. He expressed disappointment that operatives were still being used in various occupations round the cells.

'How soon will it be,' he asked, perhaps disingenuously, 'before there are no people here at all?'

We told him that we did not see that day ever happening. He was probably too polite to say so directly, but he obviously considered that we were too conservative, too risk-averse, too careful by half. We were not the kind of people who should be leading the new industrial revolution.

Yet all the messages that we were picking up from the rest of British industry and, especially, from company visits that we made to Japan was that the chance of failure in applying new technology was in direct proportion to the ambition of local engineers.

A chart constructed by Ingersoll Engineers (an engineering consultancy) was published in the *Financial Times* as early as May 1985. It projected potential technology and degree of risk as shown in the table on the following page. The higher the technology, the more risk and more chance of failure. In the top high-risk/high-technology application, the chance of things not going to plan is as high as 80%. Why?

(a) Let history judge. Far from not welcoming high-tech solutions to their problems, British manufacturers have welcomed sophisticated computer applications and have often done their best to make them work. There is scarcely a medium- to large-sized company in the UK without its management services department, plus development engineers, trying to use their local computer mainframe effectively. However, there has been far too much gullibility in believing what computer

Technology and degree of risk	Investment type
High risk/high technology	Computer-aided manufacture Computer-aided design and manufacture Flexible manufacturing systems Integrated database
Medium risk/medium technology	Computer-aided design Materials requirements planning Direct numerical control Robots
Low technology	Just-in-time Cellular Manufacturing Computer numerical control
No technology	Simplification in process Layout changes Numerical control

experts have promised—and eventually not delivered. No one should be taken in by people who propagate 'factory of the future' solutions to problems which are mainly in workforce motivation.

(b) The risk of a 'leap in the dark' to a 'lights-out operating factory' seems too great with the available skills, aptitudes and limited numbers of people who believe that they can produce a high-tech future.

The robotic cells we introduced seemed something of a gamble for the expertise we had. Permission to go ahead was given only when much of the equipment involved was to be bought directly from suppliers with a proven record.

By far the best approach appeared to be an incremental one within an overall plan. Step-by-step is much better than an attempt to produce a massive leap forward.

(c) Unless the factory culture is changed, especially in terms of work organization and payment systems, new plant alone is unlikely to give favourable results. It often needs flexibility in manning and especially in relationships between skilled and unskilled operatives which 'low-tech' equipment might not need.

(d) Engineers who have not had experience of introducing CIM or flexible manufacturing systems, or indeed any major development, make mistakes. They deserve support, but more importantly for their own confidence they need quick successes. Only the incremental approach to high-technological application will achieve this.

(e) Nothing should happen before a total manufacturing strategy has been developed, within which the technology factor can be placed, discussed and if necessary accepted, with whatever priority seems appropriate.

(f) Simplicity should be paramount in whatever is done. The Japanese appear to go for simple rather than technical solutions. So should we.

3.2 HIGH TECHNOLOGY AND THE PRODUCTION PROCESS

Early in the eighties, we took the incremental approach to heart, and looked at one part of the Chapel factory where production processes and plant layout had been developed and extended progressively over the years. In fact, it was a typical production area where successive engineering applications had been introduced, all thought to be beneficial, but the result was less than optimum:
- a tortuous product/manufacturing routing had been produced;
- facilities were often duplicated;
- there was excessive handling;
- WIP and associated finished goods stock were far too high.

We appreciated also that a large part of the complexity discovered resulted from:
- the apparent (as opposed to real) needs of customers;
- new materials and updated specifications (again apparently needed);
- the mixing of product markets in the production process. We did not dedicate machine or production areas for one product market or another. OE and replacements were mixed up.

Once the problem had been recorded, we considered carefully what investment could give the highest return. There was no possibility of having an 'act of faith' which many proponents of high technology apparently believe is necessary. We needed cost reduction, high quality output and standardized products. As well, we required:
- improved workflow
- reduced handling
- simplification in all processes
- reduced lead time
- enhanced opportunities to create a better 'culture' on the shop floor with more effective team working.

The problem of relating such potential improvements into the rigorous financial data needed for DCF investment appraisal calculation was recognized. The method of carrying out this part of the analysis was to see how far the less tangible benefits would improve the strategic fit and improve associated contribution.

To confirm the benefits to strategic fit we collected the following data:

- Work organization: number of teams and their structure; people of different categories (directs, indirects, supervisors, maintenance etc.).
- Products made: now and in the future—volume, type, material quality, process.
- Production: machines - capacities, speeds and manning; workload/capacity utilization; Actual work vs waiting time; inspection—people and time; material flow.

- Costs: Standard and actual cost per operation/process; material cost and utilization; rectification/scrap costs.
- Associated systems: planning; bonus payments; maintenance.

What possible technical and other changes could help achieve the improvements we needed? The providers of high technology solutions set out the possibilities. Their solutions all point in one direction only—the use of sophisticated, computerized planning and associated manufacturing equipment. All proposals are full of acronyms. For example:

- CIM: The Computer-Integrated Manufacturing environment, says the DTI Enterprise Initiative pamphlet, *Managing in the 90s*, is one where order processing, planning, making (production technology) and distribution are all linked together. MRP II (Manufacturing Resource Planning, of which more later in Chapter 6) is part of the process. CIM, it is said, uses information technology to improve the competitiveness of the business. Integration of all processes in a manufacturing activity is the key to making the business prosperous. The day of the stand-alone computer system is gone. CIM is a business, not just a technological, strategy. CIM it is said, will gain all the benefits that a good production manager is looking for:

— competitive production
— reduced lead time
— improved productivity
— information to manage the business
— integration of engineering and production functions.

Anyone associated with UK manufacturing companies over the last twenty years will remember the proposals to integrate systems with production activities which, despite tremendous effort and good intentions, have gone wrong. We looked very carefully at IBM's 'COPICs' computer-based system in the early seventies, but the then-available software did not warrant an active pursuit of the total system. Similarly, new software is still a problem, but so are the surrounding motivation and work organization changes which are necessary. Production managers should look at CIM with a degree of doubt if all other elements in the manufacturing strategy formulation appear to be missing.

- MAP: Manufacturing Automation Protocol is a 'standard' method of transferring data from one element in the CIM activity to another. A 'protocol' is a set of rules which allows information to be sent from computers to computer-driven equipment.
- CAD/CAM: Computer-Aided Design/Computer-Aided Manufacture is a method (or methods) of using computers to design products and associated tooling. Anything which reduces the laborious drawing office process of tool design should be useful—as indeed we found.

    Computer-aided manufacture should (in theory) provide the means for deploying CAD output onto the shop floor or tool-making room. This again we have found useful.

- FMS: Flexible Manufacturing Systems are normally defined as a group of machines which are controlled by a computer. Our robotic cells were so controlled.
- Pick-and-place units and robots: This is specialized equipment dedicated to carrying out one or two production activities. They can be (as again we found out) very useful in replacing boring, repetitive jobs which normally are carried out by manual labour. 'Picking up and putting down' is a good case in point.
- Automatic Guided Vehicles: These, it is said, can replace the myriads of fork-lift trucks and manually conveying vehicles which have traditionally formed the means of transferring production between the factory, warehouse and distribution.

It is fairly easy to design a plan or progressive chart, covering the various technologies/philosophies available. We did. It looked as the table below.

CIM	Computer-Integrated Manufacturing
JIT	Just-in-Time
SPC	Statistical Process Control
TQM	Total Quality Management
Group working/Payment/Culture/Training	
Focused Businesses	
MRP II	Manufacturing Resource Planning
MRP	
CAM	Computer-aided manufacturing
CAD	Computer-aided design
AMT	Automatic Manufacturing Technology

If we wanted to become a 'world-class manufacturer', apparently we would need to climb up the ladder until we had a complete and effective computer-integrated manufacturing system. The higher we climbed the more would long and variable lead times disappear, along with poor quality products, rising costs, inadequate customer service, capacity imbalance and high inventories and overtime—or so the engineers told us. It would only need the total commitment of everyone (unions included) and money.

To give the engineers credit, the robotic cell concept of manufacture which we eventually introduced, plus MRP II, did indeed improve performance and many of the needed benefits were achieved. This, though, was not before considerable sweat and time had been spent in improving cycle times and ensuring that all the bought-out items actually related to those designed in-house. However, a full CIM system has still to be achieved or even fully planned. Our engineers considered that the final ultimate technological goal had to be CIM—computer-integrated manufacturing.

Our analysis of CIM and what it meant to the company is covered in section 3.3.

3.3 ELEMENTS OF CIM

(a) CIM embraces 'islands of technology' such as CAD/CAM and MRP II and links them together to form one whole.
(b) It is said that the full benefits of (say) minimum stock and the best possible customer service will never be achieved unless the 'islands' are linked.
(c) The scope of CIM is shown in Fig. 3.1. All activities should be integrated and no one function should carry out its activity in isolation.
(d) The integration should take place through:

- interface computer programs
- computer systems and common databases
- team activities
- strategic/policy analysis and associated plans
- local area networks.

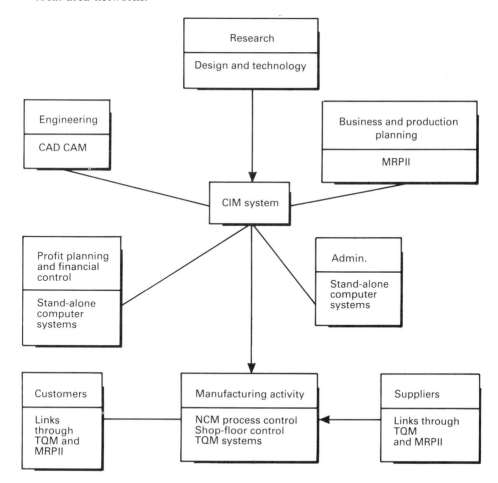

Fig. 3.1. CIM in operation.

(e) The obvious mechanism of developing CIM is through the use of a framework similar to the one outlined in this book. It is likely that generating interrelationships will be very important in achieving success.
(f) As will be discussed in Chapter 6, it is likely that a necessary precursor to CIM will be a fully implemented MRP II system especially if this can be linked up with major suppliers and customers.
(g) As far as computers are concerned, they need to be integrated with manufacturing, not just to aid the production process.
(h) The use of standard communication protocols (MAP/TOP/OSI) is totally inevitable.
(i) It is likely that computer systems hierarchy must be in place, similar to the one shown in Fig. 3.2.

3.3.1 Why moving to CIM may be difficult

There are many reasons why the road to CIM may be unclear, perhaps even dangerous. These are:

(a) The work force. It could be ill-trained, unconvinced of the need for CIM and perhaps overpaid for what it already does, and is unlikely to look easily at new techniques which will take away some of its obsolete (though pleasant) working methods.
(b) Inadequately resolved strategic fit between production and product markets. This might need considerable amendment before CIM is appropriate.
(c) Comparatively poor management, who may look upon CIM as an alternative to their own inadequacies.
(d) Investment return—too small considering the risk involved.
(e) The current 'islands of technology' and stand-alone computer systems, which appear to be working well and need to be kept as such.
(f) Impossibility of achieving zero-defects, due to changing customer specifications and the nature of incoming raw materials.
(g) A poor culture, in which CIM cannot operate properly.
(h) Poor work organization.
(i) The whole suggests a strategy where technology plays a part but not necessarily the most important part in a total manufacturing company strategy.
(j) Implementing CIM completely and effectively may take anything from 5 to 10 years. Senior management might lose interest during this time. Specialists might come and go. Creating the initiative, motivating everyone, and keeping the process rolling at a satisfactory rate could be difficult.
(k) Ignoring the expertise available outside the company—i.e., other companies, installations, consultants, the DTI, universities, etc.

3.3.2 Starting out on CIM

As in attempting to introduce MRP II, the essential precursor to getting a CIM strategy in process is to gain accurate knowledge of everything to do with processes, products, standards of performance, customers, suppliers, etc. This knowledge should

Sec. 3.3] Elements of CIM 67

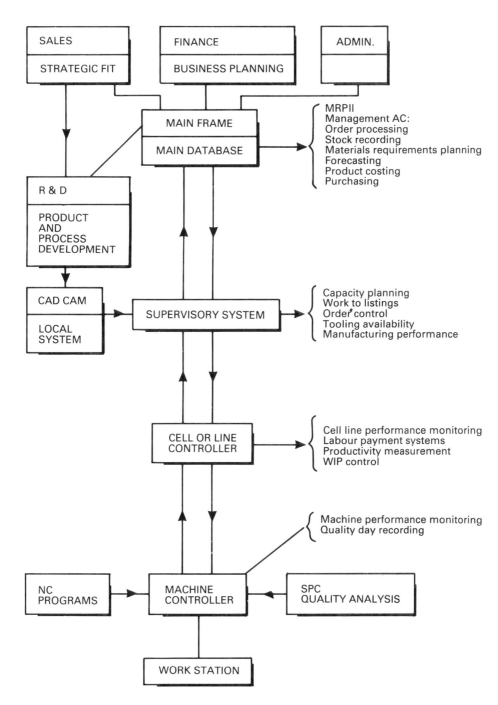

Fig. 3.2. CIM—the computer hierarchy.

form the company database. From then on, the process should follow these lines.

(a) Determine the 'Mission Statement' and the 'What Kind of Company' statement.
(b) Determine broad company objectives and the gap between current performance and these objectives.
(c) Draw up strategic-fit data sheets and determine where new technology and CIM will help.
(d) Determine manufacturing technology changes, how these are to be funded, and how quickly they can be introduced.
(e) Establish a plan which will integrate current and perhaps future 'islands of technology'. Consider all current computer equipment and systems and determine how far linking appears possible now and in the future.
(f) Probably start with MRP II and its associated database and extend outwards, linking CAD/CAM and process control.
(g) Determine how far automation or robotics is to be taken and what will still need human intervention.
(h) Start in a minor way on linking projects which can be quickly cost-justified or which perhaps do not need new and untried technologies.
(i) As in all major projects, a steering committee is essential, with project teams taking over agreed projects.
(j) Training and education at every stage is essential.

3.3.3 Benefits of CIM
CIM should operate within a broad manufacturing strategy. That said, the needs which the strategic fit analysis should highlight ought to be achieved, not least these:
- capacity and order load
- shorter and more consistent lead times
- minimum work in progress
- orders produced as and when stipulated
- order status updated rapidly
- re-scheduling flexibility improved
- real-time quality monitoring improved
- operative performance less crucial than in the past
- introduction of new products eased
- improved material utilization
- accurate reporting of plant and labour efficiencies.

3.4 ADVANCED MANUFACTURING TECHNOLOGY IN ACTION

Having expressed a degree of cynicism about the use of technology to drive manufacturing productivity higher, it is essential that examples of apparently successful applications of advanced manufacturing technology are given.

As early as June 1987, IBM produced a booklet *Computer Integrated Manufacturing—the IBM experience*. In the foreword, Sir Edwin Nixon states that 'CIM represents a clean sheet of paper for European industry. It provides the opportunity to line up on the starting-blocks alongside the most successful industrial nations of the world.'

The basis of the new technologies, according to IBM, lies in building programmable microprocessors into most machine tools, processes, robots, and handling and storage equipment. These processors in turn are part of a hierarchical organization of computers which are geared to cell, area and plant control.

A manufacturing plant should aim for continuous-flow manufacturing (CFM). All activities which do not add value to the product, like handling, should be eliminated. It does not matter whether the manufacturing process makes high or low volume products: CFM applies.

As far as computers are concerned, companies should move from computer-aided manufacture to computer-integrated manufacture as soon as possible. Most of the framework elements discussed in this book then become part of CIM:

- group technology
- balanced material flow with the process (see section 3.5)
- Kanban or JIT
- supplier integration
- zero-defects
- MRP II
- multi-skilling.

Integration becomes a key business strategy. The key to integration is information. Integrated business systems will not work properly without a systems network architecture (SNA) (in IBM's case) and an international standards organization (WOSI) which will link equipment from different suppliers.

There is a useful section in the IBM booklet, covering 'Evolution towards integration': 'CIM will grow from a variety of stages depending on the relative progress being made in the various CIM components. The main component will be an overall strategy, without which a company would just be automating its problems.'

A further booklet, *Boardroom Report: Advanced Manufacturing Technology* (Findlay Publications, November 1988) highlighted how AMT was being tackled in Hoover, Sturmey Archer, Massey Ferguson, APV Baker PMC, Nautech and Holset Engineering, Mars and JCB.

A review of the companies concerned revealed perhaps the obvious conclusion that AMT is not primarily a technology problem; it is more to do with management and company culture.

All the companies quoted report considerable benefits from AMT. Some went wrong initially owing to a wrong choice of technology and a failure to realize the impact new technology would have on the organization as a whole.

At Hobet Engineering, the problems were not basically technical, but concerned workflow, dataflow and logistics.

Much of the future in the companies described is embodied in:
- linking islands of technology
- integration to allow products to be made to order
- manufacturing lot sizes of one.

Mars management, however, admits that though there is a lot of understanding between functions, there is some way to go in terms of refinement.

3.5. MATERIAL FLOW AND TECHNOLOGY

Is there another approach to significant change and improvements in manufacturing efficiency which does not need a plethora of computers and computer technology? The advocates of optimized product technology (OPT) appear to offer a non-computer-dominated solution to many production problems.

Optimized production technology is built around material flow and the recognition of bottleneck and non-bottleneck processes. So, the first rule of OPT is that balancing flow, not maximizing capacity, is the crucial factor in improving manufacturing performance. It is not (it is said) essential to keep all machines and all operatives busy all the time, but it is imperative to keep work flowing at a predetermined rate. (The rate is set by the bottleneck operation.) If presses making a car disc brake pad are operating at speeds faster than some downstream operations can handle them, then the presses are really being misused.

Gaining time by careful scheduling of non-bottleneck operations is not fruitful. Scheduling a bottleneck normally will be. The bottleneck will condition/determine total output, lead times and eventually levels of work-in-progress and finished goods stock. It would be useful, therefore, to identify the bottleneck and determine its potential earning power in terms of contribution per hour. Local operatives, first-line supervisors and schedulers should all then know that if the bottleneck operation ceases to function then 'X' contribution per hour will be lost. That kind of financial evaluation should be an important part in the analysis before any new technology is introduced.

So perhaps the proponents of OPT should rewrite their philosophical rules as follows:
- identify bottlenecks in the process
- keep all bottlenecks as busy as possible
  schedule work content to make sure that other operations do not become bottlenecks
  tolerate idle time on non-bottleneck activities
- use production technology to eliminate all bottlenecks
- do not use high-powered computer technology to optimize non-bottleneck activities
  do not attempt to maximize the output of the factory in total
  an hour saved on a bottleneck operation is an hour won for the whole process.

3.6 THE FERODO APPROACH

Seventeen years separates two articles published in the *Financial Times*, each of which describes technological developments associated with production in Ferodo.

The first, dated 2 January 1970 records how a mixture of Research and Development specialists, plus academics from the University of Science and Technology in Manchester, had introduced 'Group Technology', into the Chapel factory.

The heading of the article, 'How Ferodo broke its delivery problems', could have provoked some bitter comments at the end of the seventies, when very little came off the factory on time and what did emerge was extremely costly. Unfortunately, most people had forgotten about 'Group Technology' by then, especially its principles.

The Group Technology concept should have worked. Why did it fail?

(a) Research and Development personnel plus UMIST academics took over 'all technical problems, plant designs, plant layout, O&M', leaving the works to cope with industrial relations and general production. In other words, this was to be a specialists' activity taking over the production process yet leaving some activities like 'industrial relations' outside the activity.

(b) The people involved identified 30 000 part numbers. These had in fact been allowed to grow to 46 000 by the end of the decade. A major effort might have been made to investigate and reduce the range (which was eventually done); instead, it was anticipated that Group Technology could handle the number of items found.

The article stated that product-costing of the 30 000 items was impossible. If it had been possible, then the amount of contribution per piece and per operation (and especially the bottleneck operations) might have been determined and some form of OPT adopted. This did not happen, though subsequently it was proved to be possible.

(c) An improved shop-scheduling (such as OPT offers) could have been used to build up knowledge of product-work content and potential bottlenecks. Additional plant (comparatively cheap at that time) might have been installed.

(d) It was expected that operatives would move from job to job to provide flexibility. In fact, this rarely occurred. Operatives preferred one job to another. They became skilled at carrying out one operation and disliked moving. Work organizations and associated payment schemes were not developed sufficiently. Team-working with all its ramifications was not introduced.

(e) The key to the failure must however be that a total change was needed in product markets, resource utilization, work organization, payment systems and in systems generally, as well as in technology.

Seventeen years later, Ferodo technology again featured in the *Financial Times* (on 18 March 1987). This time the heading was slightly different—'No brake on the pace of change.' It is significant that this article quoted objectives for the new technology which were precisely the same as those which Group Technology was thought to gain:

Increased quality standards
Reduced costs
Guaranteed output rates
Guaranteed delivery dates
Reduced work-in-progress and decreased lead times
Ability to handle variety and complexity
Elimination of human error
Improved material utilization.

The solution to gaining these objectives appeared to our engineers to be in[†]

- Computer-aided drafting (CAD) for tool design
- Computer-aided manufacture for tool manufacture
- The provision of computerized, automatic material weighing (a crucial factor ensuring that material input to production would provide required quality products)
- A production system based on independent cells, each capable of forming and finishing a variety of products
- Linked finishing operations which would largely be automatic
- Each cell to be controlled by a microprocessor unit interpreting the day-by-day workload, monitoring cycle times and output generally
- Teams comprising mixed union representation to carry out the operations without any incentive payments whatsoever
- An incremental approach to be made proving the technology on a step-by-step basis, ensuring that the investment paid off from the moment the new equipment was installed.

Total strategy
Fig. 3.3 sets out the initial manufacturing strategy undertaken. We had no thought of introducing computerized integrated manufacturing, only to ensure that the strategy as a whole held together. We were determined that:

(a) the investment would work;
(b) the unions and work force would be positively cooperative;
(c) there would be an immediate pay-off in quality improvement and product market service;
(d) focused factories/product lines would provide a strong motivational basis for ensuring that the overall strategy would work;
(e) the introduction of new technology would not be dominated by engineers but be a company-wide concern.

Computer use
Fig. 3.4 shows the simple strategy planned to deploy computers in the activity. The mainframe was needed to run the MRP II package. The supervisory system could take the MPS run and institute capacity planning and scheduling. Cell controllers were needed to monitor local performance.

Results
All went well until the plan for computerizing the finishing operations was reconsidered. It was found that the majority of savings envisaged for the total project had already been achieved by instituting cellular robotic production.

[†] As quoted in the Ferodo Annual Report for 1984.

Item	Objectives	Linking
Product/ market	Ensure expanding revenue/ contribution	Technology, part of strategic fit, as well as SQA
SQA	Reduce rejects, improve quality. Reduce people involved only with inspection	Essential part of product market activity. Vital to gain new OE contacts
Motivation/ payment	Aid improvements in productivity on a year-by-year basis. Eliminate relative deprivation and 'red circle' jobs when payment systems have gone wrong	Acceptance of new technology and required flexibility, training the team working to achieve maximum use of new equipment
Focused factories Work organization Training	Improve profit/ contribution and effectiveness of management at all levels. Improved resource utilization	Essential in achieving improved motivation and payment systems and effective manning of new technology
Robotic cell manufacturing	Reduced costs. Lower WIP. Better quality. Improved service	Essential to make product-market objectives, especially in flexibility of response to market demand
MRP II Resources	Improved resource utilization at every stage in order processing materials, plant, people, working capital	Essential in linking functions and responsibilities throughout the company, heading for CIM, if not actually achieving it
CAD/CAM	Reduce development time. Enhance tool design (better tooling). Reduce costs	Improved flexibility. Much better service to robotic cells and product markets, ensuring speedy introduction of new products.

Fig. 3.3. Ferodo strategy—linking processes.

A sound rule therefore in introducing new technology is not just to consider the complete application and determine potential gain, but to break the project into discrete component modules (if this is possible) and consider them as separate investments.

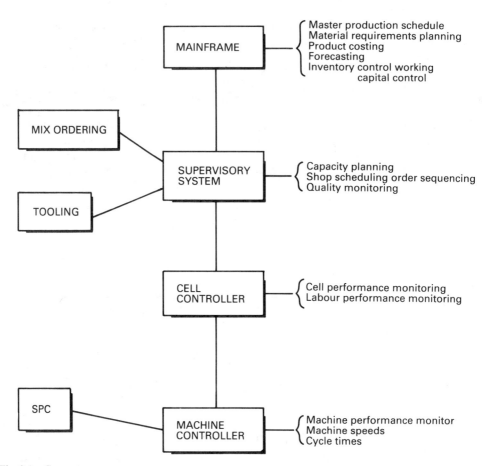

Fig. 3.4. Computer usage.

The finishing-stage computerization part of the project was made more difficult because:

- stability had not been totally achieved in the product range nor its material make-up;
- flexibility was demanded of the cells, before the organizational changes to help achieve it had been agreed and put in place;
- payment systems had to be changed, because the rest of the factory had not been brought into line quickly enough;
  demand fluctuated and product variability actually increased, largely due to the slow application of MRP II.

All this was proof that having a manufacturing strategy covering all aspects of the business was not enough if the component parts did not come on stream when needed. Even so, one production finishing line was linked to the manufacturing cells, with a judicious use of pick-and-place units, programmed to accept any variety of finished product and eliminate most manual interventions.

Full automation of the rest of the processes was not feasible, mainly because the technology proved far more inflexible in handling product variety than had been envisaged in either computer simulation or in physical trials. Some manual intervention was still needed. Introducing high-tech plants tends to create, rather than eliminate, constraints in the production process.

The application of new technology on a fairly wide scale brought opportunities to change surrounding systems and organizations, job structures and titles; and, for the first time, teams of MSF, CSEU and T&GWU people worked together. Regrettably, flexibility between jobs held by different unions was not as extensive as it could have been.

Opportunities come irregularly to make fundamental changes in culture and work organization: new technology was one such occasion.

A programme for the nineties
Like many other companies, Ferodo ended the eighties with a series of islands of technology. Fig. 3.5 gives a broad indication of them.

(a) One of the first considerations for development is to determine how far the islands need to be linked together.
(b) Computers are becoming more and more intelligent and cheap. Their use must expand, but within the broad systems/computer development shown in Fig. 3.4.
(c) All company functions need to be integrated in some way one with another. Without such integration, inventories and customer service will only be improved marginally, costs will remain high, there will be no continuous improvement. The aim should be to have a computer-integrated, not computer-directed process.
(d) The key to integration will be in having information which everyone can access.
(e) All the traditional methods of manufacturing, e.g. using fork-lift trucks that help overcome poor plant layout, stocks that hide poor scheduling and poor supervision, should go.
(f) Irrespective of whether a 'Brown-field site' has now been achieved, it must operate as if it were a green-field one.
(g) Achieve zero-defects even though there are constant changes to product specifications.
(h) Ensure that using CIM or AMT makes a significant impact on production costs. A buyer from an American company once offered me a considerable business contract. The Japanese, he said, had offered to make the contract at 'X' price and guaranteed, through their ability to make constant improvements, to reduce the price by 3% per year for the next 5 years. Could we match it? Reluctantly I said, 'No'. Survival in the nineties must produce a competition situation where such contracts can be taken.

76 Technology [Ch. 3

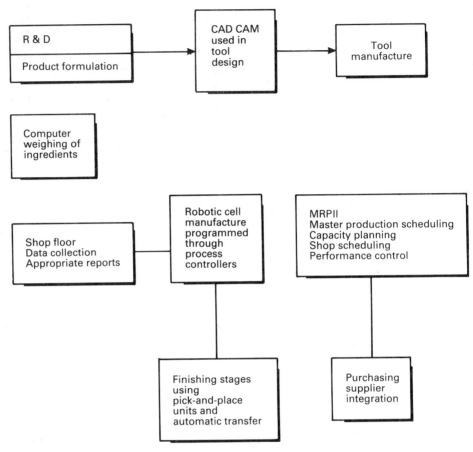

Fig. 3.5. Islands of technology.

(i) The workforce, therefore, must be reduced consistently. The people who remain must be better trained and educated generally. The average shop-floor worker is currently not good enough.
(j) Wherever possible, still use comparatively low-technology equipment, which helps to re-design work flow.
(k) Outside support will be needed for:
 — hardware and software control
 — communication equipment
 — data management
 — applications
 — user support
 — business function management.
(l) The whole process will still be evolutionary within the broad business and technology strategy agreed.

(m) No process should be introduced which does not use flow or cell manufacturing techniques or a combination of both.
(n) Production engineers should concentrate on reducing time spent on tool change-overs and sets to an absolute minimum.
(o) Operatives to be multiskilled in maintenance, tool changing and general production.
(p) Most products should be made as often as possible using minimum batch sizes.
(q) As much new equipment as possible should be bought, after it has been proven. Little should be made in-house.
(r) A first integration exercise should link MRP II with the production process. This should help to reduce set-ups/tool changes, ensure that dimensions are achieved, monitor capacity usage and output, and drive production equipment.
(s) GT and OPT technology should operate wherever possible.

3.7 TECHNOLOGY AND THE MONEY PROBLEM

For a company long starved of capital investment, our problem was not what to spend money on, but what could possibly be left out. Priorities abounded.

With limited engineering experience and expertise in installing sophisticated plant, the pace of change must always be limited unless outside help is sought. For us, and surely for most companies, there is a finite limit on the amount of cash available for new investment. There always will be. So, even if the pace of change is considered to be slow, there were and always will be reasons for this.

A very useful agreement with Lord Tombs was that we could anticipate spending half the operating profit we generated on capital equipment. (Gradually the proportion was allowed to increase.) This agreement gave an extra dimension to the need to achieve high operating profit. It was also useful in debating profit with our unions. We could show them a direct relationship between profit earned and money spent on securing the future of the company.

It was also useful for the engineers to know that current profit performance could produce future investment. They probably worked harder to achieve earlier success than perhaps they might have done otherwise.

The finite amount of money made available motivated a careful debate on how it should be spent. No money was available other than that which we had already earned—a useful internal discipline.

The priorities for investment will vary from company to company, but in most cases they will be:

(a) Health and safety concerns, particularly those where our own workforce and local inhabitants could be at risk. The COSHH regulations were not unduly concerning, because we had already pre-empted them.
(b) Structural repairs—to the site and buildings.
(c) Cars—a major bone of contention, but one which most managers will not let go.
(d) Investment based on achieving required investment return.
(e) Other investment.

Included in (d) was equipment needed to increase production capacity, improve efficiency and reduce cost. If these factors could be combined, so much the better. Within the broad, technical/manufacturing strategic plan, engineers had to bid for their own capital. A major problem remained. Did the traditional discounted cash flow method of appraising capital investment still apply? Was DCF still relevant?

The Japanese, it is said, ask a different question when confronted by potential capital expenditure—'Will my company grow as a result of this investment?'

Our own investment was certainly conditioned by current interest rates, even though the money was self-generated. If the alternative to putting more robotic manufacturing cells into the factory and obtaining a 20% return is putting money on deposit with no risk at 15%, then investing in new plant could be unjustifiable. The West Germans with interest rates for many years of around half those of the UK, can equally justifiably invest more with lower risk in plant and equipment.

We continued to use the DCF appraisal method because:

(a) Everyone understood it. It could give some poor results and some vital equipment might not be obtained, but the financial discipline it imposed always seemed to be more important than the possibility of lost opportunities.
(b) If reasonable intangible benefits are also taken into account, such as a reduction in work-in-progress, at, say a 12% rate of interest, then the system can become slightly more flexible.
(c) Any alternative to DCF seemed to give a potentially dangerous result. Backing 'acts of faith' seemed particularly wrong.
(d) At the same time that we discussed the possibilities of alternatives to DCF, we started to increase depreciation rates to ensure that plant was replaced more frequently than at any time in the past. If greater risk is entailed in investment, then depreciation rates should reflect the extra risk.
(e) DCF investment appraisals and subsequent evaluation of results have an important impact on the potential success of introducing technology. We found that if a project does not generate its discounted cash flow in the first year then it is practically impossible to catch up in subsequent years. Our engineers knew this and it was always a useful spur to get new plant running as quickly as possible at planned cost.

3.8 HIGH TECHNOLOGY—THE ONLY ANSWER TO MANUFACTURING PROBLEMS?

In November 1989, the journal *Management Today* published an article entitled 'Britain's Best Factories' and described five of them. All had introduced CIM/JIT/TQM/MRP II, etc., with the inference that this was the way, not just to equal the Japanese, but to surpass them.

Is this really the case? If it really is so, then all British industry needs to do is invest in larger and larger doses of computerized equipment.

In the UK, it has been reported† that £323 billion has recently been spent 'on robots, computers and the like; so if capital equipment alone was the solution, manufacturing companies would all have competitive edges as sharp as sabres'.

Between 1980 and 1987 General Motors in the USA‡ spent $42 billion re-equipping old factories and building new ones. Even so, GM's market share continues to decline and so does its return on capital employed and overall profitability. One of the most automated of GM's plants at Hamtramck, Michigan, with 300 robots, has lower productivity and poorer quality control than a similar, but labour-intensive, unit at Freemont, California.

A study carried out by the Massachusetts Institute of Technology‡ on automation in car plants in Europe, USA, South Korea, Taiwan and Japan concludes that levels of technology have little to do with productivity. The Japanese plants are apparently better because their workers have adjusted better to automation. Management style has encouraged workers to take responsibility for much that they do.

Even in the careful, cautious plan we made, the first batch of robots had to be changed owing to an inadequate specification on our part. Then, with the new equipment, it took many months of careful assessment of the component functions to ensure that cycle times were optimized.

The teams we had were hand-picked for the robotic cell manning, and every possible effort was made to ensure that they worked well, with a day rate basis of payment. Even so, the relative deprivation they suffered, from seeing wage-drift at work in other unreformed payment-by-results systems, caused a work stoppage.

We failed to obtain the requisite software for tool changes, which played a vital part in ensuring we had flexibility. But even if the technology works well, there is still no short cut to manufacturing efficiency. Poor managers are not suddenly transformed into brilliant exponents of the art by being given computers and computer-driven equipment. Before high-technology equipment works effectively, it is likely that good management will already be looking for something which will enable them to improve on their already good performance. New technology will be seen as another, additional method for this, not the sole, single solution.

Perhaps Nissan in Sunderland have the right answer when they say, 'Stop worrying about progress through technology and start thinking about change and improvement through people. It will get you more and get you further.'* Ferodo's response would be that you need both—technology and good work organization. One without the other will not drive the company forward in the way that it needs.

3.9 WORLD-CLASS MANUFACTURERS

Much has been written about being a 'world-class manufacturer' and about 'searches for excellence'. Does it make sense to believe that 'world-class' is a relevant objective?

---

† See Boardroom Report. *Advanced Manufacturing Technology* November 1988. Findlay Publications Ltd.
‡ See *The Economist*, May 1988.
* Told to visiting Ferodo engineers.

PA Consultants produced a summary of a seminar they held in 1985, where the priorities a company should follow if they are to become world-class manufacturers were recorded:

(a) A total preoccupation with quality. Total quality management must be practised.
(b) Everyone, especially the Chief Executive, must be committed to success.
(c) Proven technology is applied wherever possible.
(d) Considerable effort is made to plan the organization.
(e) The supply chain must be integrated with the company's operation.
(f) Organizational boundaries should not be limited—if they exist at all. Organization is flexible.
(g) Computer integrated manufacture, total quality management, and just-in-time systems are central pillars in achieving world-class manufacturing status.

CONTENTION

(1) Introducing highly expensive sophisticated technology into an already established production unit without first considering the total manufacturing system will inevitably produce disappointment.
(2) There are now enough examples in Western manufacturing industry, such as General Motors, which prove conclusively that applying high-technology solutions to production problems, far from improving efficiency and cost, could be counterproductive.
(3) If the use of CIM/TQM/JIT, etc. will produce an effective production unit and these are fairly easy to apply by everyone including competitors, then to gain competitive advantage, something else needs to be done.
(4) The use of high-technology plant is not an alternative for poor management. If management is poor, then the use of CIM/AMT, etc. will only produce further poor results.
(5) Analysing and improving material flow, concentrating first on bottleneck operations, is a useful starting point on the road to using high-technology equipment.
(6) Leaps in the dark will normally land in big holes. Most cautious managers will base their technical strategy around an incremental approach, especially in any robotic cell manufacturing technology. Integrated manufacturing could result, but comparatively slowly.

A suitable set of strategies

(1) Robotic cell manufacturing has produced favourable results, good quality products and a high return on investment. We will continue to use this technology where it gives a return of at least 30%. Applications will be carried out on an incremental basis.
(2) Improving shop-floor efficiency on a year-by-year basis is essential. Technology should play a part in this activity, but only within the strict financial parameters which are acceptable. Numbers employed on the shop floor should fall compared with the output achieved.

(3) Providing the strategic fit with the relevant product market will be a key element in deciding whether and how fast we progress towards CIM.
(4) Linking processes, eliminating work-in-progress, ensuring production lines are able to operate more flexibly than in the past will help decide technological/investment solutions.
(5) We will continue to buy out most of the high-tech equipment we need. We no longer have the internal expertise to design our own equipment. Outside purchases should be of well-proven equipment, from suppliers who provide excellent back-up services.
(6) As far as possible we will install capacity which will provide greater output than that apparently needed, going by current forecasts.

4

Resource utilization and cost control

4.1 INTRODUCTION

In order to survive in late 1982, spending in Ferodo was so reduced that only the purchase of raw materials and spares to keep production equipment operational, plus the payment of the people still employed, was allowed. The Purchase Department was given a sum of money which could be spent each month. Once it was used up, no more was made available. Pay cuts were imposed and even senior managers were put onto short-time working.

For nearly a year overtime ceased. Expenses shrank to minuscule proportions. Visits to other companies and organizations and even to customers were curtailed. Company pool cars, always a good indication of attitudes towards spending, sat in the garage, forlorn and neglected. Power was switched off when not needed for production purposes. Temperatures in offices were allowed to fall below levels which most people once thought much too cold. Unheard-of savings were made in every department and function.

The outcome was a positive cash generation, never previously achieved in the history of the company. It was not a regime that we would want to maintain for very long, but it proved conclusively that a reduction of costs of between 20% and 25% was possible, while still keeping the company effective.

Sometime in 1980, a decree had been issued to reduce staff arbitrarily by 15%. Many believed that there was no overmanning and that all employees were doing essential jobs. Reluctantly the cuts were made. Subsequently the company continued as if nothing had happened.

Later, people reductions became more difficult. The holes left had to be filled judiciously, but there was (and still is) continuing proof that some people reductions are always possible.

The UK motor industry and, particularly in recent times, the component parts of that industry have become singularly adept at providing occasions and opportunities for a degree of self-indulgence. These may not be totally wrong, but they tend to undermine a tough, cost-conscious, performance-orientated organization. All manufacturing companies need to regard spending money as anathema if there appears to be no tangible reward for doing so.

The Board of Directors and senior managers need not necessarily be seen as Cromwellian, but at least they should all act as if they have a closed and very shallow purse. All should behave in the same way. Having one function spend money easily while the rest control it tightly should never be allowed.

BTR plc sets a good example. John Cahill[†], quoted in the *Independent* of 2 April 1990, says:

> We have dedicated ourselves in the past five years to getting our costs down and have invested over one billion pounds to achieve that objective. About 28% of the increase in BTR's profit last year was a direct result of that programme.

The BTR philosophy is an attractive one. It attempts to maximize profit in niche markets by cost reduction, rather than by chasing market share.

The Japanese also give a good lead. At the 'Motor Industry in Britain—which way in the nineties?' seminar of 19 November 1988, it was reported that, despite the current high performance of Japanese car manufacturers, they are still cutting costs determinedly:

Mazda — The 'drill operation' rationalization programme reviews every facet of parts manufacture and assembly processes.
Toyota — 'C50' campaign is aimed at halving administration cost.
Nissan — 'Catchwork Scrooge' is aimed at halving administration costs.
Honda — The 3.3.3 scheme calls for greater financial effectiveness through a fall of a third in labour force and buying costs and a 33% boost in automation.

Resources which can be deployed within a manufacturing strategy are always too few. Spending unnecessarily is not therefore an option for anyone.

4.2 PEOPLE

People are by far the most costly resource used by a company, especially a motor components company. Even if direct salaries or other on-costs are reasonable, it is the additional cost burden that people impose that is significant. Because someone is employed to carry out a function, cost is incurred as a result, e.g. the bored R&D specialist who convinces his manager that it is vital that he flies to West Germany to discuss a project with a potential supplier; the representative who does not actually cheat on his travelling expenses, but who always makes sure that his expense account is nice and full; the secretary who thinks she needs a personal computer, even though a simple basic word processor would do; or the clerk who phones up his cousin in Melbourne to ask about the health of Uncle Jim.

[†] Then Managing Director of BTR plc.

(a) *Technology and people.* The writers on CIM/AMT/JIT, etc. might have it wrong when they report that direct labour can be ignored as a cost because it is only 7% or even 10% of revenue. It is not the cost that should be of concern but the loss or gain in contribution and revenue which should be measured and improved. Direct labour needs to be ever more productive.

(b) *Reductions in people.* Computers have singularly failed to effect any major reduction in numbers employed. Old-fashioned techniques such as work simplification have been proved to produce far greater reductions. Section 4.4(d) sets out how this might be achieved.

(c) *Manpower planning/succession planning.* One of the most dispiriting reviews an MD can make is to analyse the current strengths and weaknesses of his management team and attempt to compare their intelligence, loyalty and knowledge of their jobs with the team of ten years ago.

The early eighties were a bad time in which to improve ability by recruitment. Many managers left or were forced to go. Few were recruited. It became much easier to promote those who were available internally rather than go outside and get better people.

Effective manpower succession planning is practically impossible. It is probably tedious and unnecessary to fill out neat organization charts with possible successors to those who are already in place. People leave, die, become ineffective. The whole business can become corrupt. It is better by far to ensure that:

- Training is effective and relevant, mainly as set out in Chapter 5
- Out of, say, a management cadre of 150, at least ten should be of Board potential.
- Recruitment at the most junior management level is consistent and good. There is no excuse not to recruit reasonably well qualified and potentially effective people.
- Career progression should be seen to work well.
- Young people should be given an early opportunity to perform key organizational tasks.
- The management team, through recruitment, retirement and leaving, should change on average by at least 10% a year. There should always be two good candidates for a management vacancy.

(d) *Redundancy.* Between 1 January 1980 and 1 January 1983, Ferodo reduced its total work force by 35%. This was the common lot of many manufacturing companies in the UK in those years. We were not unique. Redundancy has scarcely stopped since.

For many people in their late fifties, early retirement, whether expressed as redundancy or not, can be very pleasant. Pensions are reasonable, redundancy pay fairly generous, the possibility of some other paid occupation coming along is quite high. Even in the best-run companies, redundancy must always be a possibility.

Voluntary redundancy is always preferable to that forced onto the organization. Strict rules should apply, e.g. volunteers can go if their job can truly be made redundant, although a useful alternative is that redundancy can still apply if the

job can be absorbed by other personnel, and if more than half the cost savings can accrue to the company.
(e) *Demographic trends.* One of the most potent forces for change in the 1990s is the demographic trend and especially the potential dearth of teenaged recruits. While redundancy could be regarded with favour at one end of the age-scale, teenaged replacements will be harder to come by at the other. However, fewer teenagers will not necessarily mean better educated ones. The belief that all managers need to have a degree may be shaken.

All companies must be 'up-skilled' with less brawn and more brain. How this is to be done needs careful analysis within current demographic trends.

The possibilities are:

- Develop the 'train principle' further, to ensure that the company prospers, almost irrespective of the average intelligence of managers.
- Reward junior managers with good qualifications better than those without, ignoring union or other protestations.
- Recruit educated mothers who believe they are not going to have more children. Employ flexible working patterns to attract them.
- Recruit more women generally.
- Recruit Third World people.
- Buy in specialist services on a short-term basis if all else fails, especially R&D talent and computer/software development.
- Subcontract activities previously carried out in-house.
- Introduce work organizations which break down hierarchies and provide more satisfactory job activities.
- Increasingly use technology to replace people on the production line, rather
  - than trying to increase their efficiency.
- Train and re-train constantly, to make average managers manage better.

4.3 CASH

If Lord Tombs imprinted anything on our minds, it was the importance of cash. Profit may come through operating the company well. It may also come from manipulating stock values or some other purely accounting/financial adjustment. Neither of these may run counter to standards of accounting practice, but they may hide or temporarily hold up the slide towards disaster. It is cash that is the ultimate measurement of how well a company is using its resources. Once this factor is forgotten or ignored, disaster comes stalking round the corner.

No company should be in trouble with bank loans if they can possibly avoid it, and good control over cash is one way to do this. Banks need to be kept at arm's length. The longer the arm the better. However, it is of little value ending the financial year with a positive cash position when a negative one was occurring for most of the year. Cash needs to be positive for most of the time if cruel interest rates and the threat of banks putting out the lights are to be avoided.

4.3.1 Cash forecasting

Making a meaningful manual cash forecast for most medium-sized companies is quite difficult. The smaller the company, the more likely that manual forecasts are possible. Getting them right is another matter. If a major customer suddenly decides not to pay, then any cash forecast can go haywire.

In medium-sized companies, monthly forecasts might be appropriate if the local banks are accommodating. But for smaller companies weekly, even daily, forecasts are essential.

How cash forecasting needs to be done was set down in our user specification for MRP II as follows:

Potential cash generation elements
    Operating profit
    Royalties
    Depreciation
    Working capital provisions
    Exceptional items including any disposal of fixed assets

Potential cash absorbing elements
    Gross stocks
    Gross trade debtors
    Trade creditors.

In both sections, items can be positive or negative ('profit' could be a loss, or stocks could reduce, etc.) Another factor to be taken into account should be gross capital expenditure less grants received, plus interest payments.

The difference between cash generation and cash usage will be net cash flow. It is likely that how operating profit has been derived will need to be shown in detail, particularly payroll costs.

The computer model we devised had these elements:

Outputs

- Future sales revenue—by product group
               market
               major customer.
  This forecast had to take into account the terms of payment and current debtor days being achieved for those markets and customers who would generate revenue in the financial period under review.

- Forecast stock values—raw materials
               work-in-progress
               finished goods stock.
  Raw materials were to be calculated by taking into account the material qualities/batches needed to make the products listed via the Operational Planning Sheets.

Work-in-progress was to be calculated by using the programming scheduling (or zone control loading) system. Valuing work-in-progress should be done by using standards established as the cost file.

- Other factors needed were:
  – Reject rates—how much extra material is needed to be bought to cover anticipated material losses.
  – Current stock levels—free and allocated.
  – Buying lead time.

4.4 REDUCTION IN ADMINISTRATION COSTS

If computers have proved something of a confidence trick in their perceived role of reducing people, what methodologies can be effectively deployed for this purpose?

Determining whether administration or clerical work is yielding value for money can be something of a problem. Applying value administration (the clerical equivalent of value analysis used in manufacturing) is far from easy. One useful starting ratio might be to calculate what added value the organization as a whole is earning and then relate departmental/functional clerical costs to it. In some functions, say, Home Sales, revenue achieved per pound spent in the last five years might be more appropriate.

(a) *Data needed.* The following measurements are needed:

- $\dfrac{\text{Added value}}{\text{All administration cost}}$ — trends for the last 5 years.
- $\dfrac{\text{Revenue earned}}{\text{Departmental cost}}$ — " " "
- $\dfrac{\text{Revenue earned}}{\text{Functional cost}}$ — " " "
- Functional cost increases over the last 5 years.
- Cost of processing one order/order line per £1 revenue over the last 5 years.
- Cost of processing one invoice per £1 revenue over the last 5 years.
- Cost of processing one production schedule per £1 revenue over the last 5 years.
- Cost of processing one purchase order per £1 revenue over the last 5 years.

A useful indicator of work-load is to determine (say) order lines per person working in the export department.

Crude though these measurements appear to be, they can indicate where costs are drifting out of control and, perhaps more importantly still, what value is being obtained for money spent. If, for example, the export department costs 40% of all sales administration expense yet only achieves 15% of revenue, it certainly needs investigating. Equally, if order processing costs take 5% of all revenue earned, this too seems expensive.

(b) *Zero budgeting.* We tried zero budgeting—a technique whereby it is assumed that all administrative departments have been scrapped and need to be re-built from scratch. From this theoretical starting point, managers bid for resources to re-start their departmental functions.

For most Ferodo managers, this turned out to be too theoretical. The inductive leap from assuming there was nothing to having a reasonable resource was too great. The built-in inertia towards bureaucracy stifled constructive thought.

We needed a much sharper approach. For example, we scrapped all computer print-out/tabulations that the mainframe produced. Then if anyone wanted one, they had to justify it to the Finance Director. In our cases this reduced tabulations by over a half.

A third, more dictatorial method still, is to say that order processing should cost no more than x pounds per order as this is all we can afford. The result can be a major cost reduction.

(c) *Functional analysis.* In many, perhaps most, organizations similar functions are carried out in different departments. For example, wage calculations and derivative data are often made by first-line supervision, plus wages department, personnel department, work study and management accounting. They should be done once only by one function.

When we set up the Operational Planning Department or Materials Control Department, purchasing, capacity planning, production requirements planning, production scheduling, performance monitoring, stock control and general order processing were all carried out as an integrated departmental function. This helped considerably to ensure that the factories were planned effectively, service to customers improved, and stock reduced; but the number of people carrying out the listed functions separately was reduced substantially.

(d) *Work simplification.* People who were once trained in O&M techniques will remember that much emphasis was placed on work simplification. This, it was said, was a participative activity where clerks would help to record what they were doing and then assist in coming up with more efficient alternatives. This rather naive approach seldom worked well. Job holders were not particularly keen on making themselves more work.

Somewhere down the years, this very relevant activity disappeared, covered up by the misguided belief that only computers could improve administration in a significant way.

We carried out work simplification in the old traditional way:

- Record what is happening.
  Log sheets filled in by participants are one way of doing this, if a firm guarantee is made that there will be no forced redundancies (a guarantee to be made rarely). Generally, a document count is fairly easy to do, even without the help of local personnel. Some measure of how long an activity should take is possible by simulation.

- Analyse what has been found.
  Duplication, work not really required, badly organized activities, might all be uncovered.

- Carry out timing.

Work measurement has always been something of a *bête noire* among clerical people. It was always correct to do it on the shop floor, never in offices, where machines dominate clerical activities more and more. Sampling is a very useful technique to discover overall activity.
- Design a better method.

  The technique suggests that if combination, elimination, etc. are applied, a new method comes out nearly automatically. It is not quite so easy, but it is nearly always possible, to combine jobs or eliminate an activity completely and then stand aside and see what repercussions—if any—occur. Usually there are none.

(e) *JIT in the office.* Later, some reservations about JIT generally will be made; but, in any case, administration should use the same philosophy as that used on the shop floor. In the early eighties, it was taking four weeks in Ferodo to process an order from receipt until it was actually launched on the shop floor. Then it took another week of paperwork preparation once the order had been made, before it could be despatched. This is a good way to ensure corporate extinction.

Response needs to be in hours not days. Once this has been accepted by everyone, the whole basis of administration changes—items of organization, attitudes, people's jobs.

4.5 PRIVATIZATION OF FIXED COST

Most organizations carry out far too many ancillary activities which are not really crucial to the mainline production function. Transport, garage facilities, tool-making, the canteen, cleaning, waste disposal, security, all come into this category.

Once the highest possible proportion of maintenance and general engineering personnel have been allocated directly to production units, it is surprising how many people are still left carrying out tool-making, building repairs, joinery, painting and decorating, and so on. Engineering people seem extremely adept at extending their functions into all sorts of activities.

It also is surprising how often these residual people tend to become disaffected. They do not have their pay linked with that of the production people. They do not get as much overtime. Their skills are not sufficiently rewarded.

Where some misguided factory manager has yielded to pressure and put, say, the waste disposal gang on a bonus linked with output, things can go badly wrong. As output increases and rejects lessen, the waste disposal people have less to do but get paid more and more.

It is useful to look at all activities not directly concerned with production, as follows:

(a) Identify all activities which could come under the 'ancillaries' banner' and have their own separate identity and perhaps autonomous working group. These could include those already mentioned above plus many others such as:

   Roof cleaning
   Box-making for export shipping
   Gardening

Internal telephone repairs
Vending machine cleaning and re-filling
Electricity substation and emergency power generation
Pallet repairing.
(b) Determine what outside/non-company organizations might provide the service. Agree a price and service levels.
(c) Discuss the results with the appropriate union groups, suggesting that personnel already employed in the function being investigated might like to bid for the contract.
(d) Consider what might happen to the people currently employed if their jobs were to be privatized. Will they be employed by the service organization? If not, are they to be made redundant? How much will this cost?
(e) Give the privatization contract to the local union group wherever possible, even if this means paying (slightly) more than an outside contractor might accept to do the job.

There are other factors involved in this operation apart from minimizing costs which need to be taken into account:

(f) Union hassle.
(g) Loyalty of the workforce concerned.
(h) Possibility that the service offered would deteriorate into something less than satisfactory.
(i) Having non-company personnel on site increases the danger of theft, and means the loss of a secure activity.

4.6 FACTORY FIXED COST

It is likely that factory fixed cost could be as high as 20% of revenue achieved. While production and engineering managers may complain vociferously of sales, distribution and administration costs, they have it within their own function to make effective use of their considerable apparently fixed cost (or reduce it).

Our approach to reducing factory fixed cost was:

(a) Cost analysis—no cost should have risen faster than revenue/added value/contribution/retail price index.
(b) As many costs as possible should be directly related to production and be the responsibility of local production management. Incentives based on cost/added value achieved should be installed.
(c) Cost comparisons with bought-in services should be made frequently and union groups made aware of significant differences.
(d) Top-down planning produces a total amount of factory fixed cost that can be spent. Managers are then asked to bid for what they think is a meaningful amount for their function—and agree it with their peers. (This in our case was perhaps the most useful way of keeping costs down.)

Sec. 4.7]   Working capital   91

(e) For specific items like energy, outside consultants were asked to carry out a survey to see what more could be saved.

Central maintenance/tool room/general services incentives were considered and not used, owing to the inability of work study practitioners to design an effective system.

4.7 WORKING CAPITAL

Working capital, as the text books say, is all the capital required for the day-to-day running of the business—cash-in-hand, trade and other debtors, raw materials, work-in-progress, finished goods stocks, etc. By the late 1980s we had whittled down our working capital to a value of around 14% of sales. From the enormous figures of seven or eight years previously, this was a major improvement.

Velocity of working capital is important, i.e. the speed at which one pound goes from receipt to debt. The velocity is delayed by the amount of working capital carried by the company. Like many companies, we attempted to offset working capital by increasing the value of creditors—by rarely paying our debts on time. How ethical this was is debatable.

Working capital is closely allied to cash and cash control. Fig. 4.1 shows a method of combining the two factors. Cash flow is shown in all its detailed elements. Movements of working capital are shown on the right-hand side of the form. If cash is not being generated at a planned rate it is quite easy to see the cause and, hopefully, correct it.

4.7.1 Stock/WIP

Consultants have attempted for many years to design and introduce inventory control systems which would reduce stock to reasonable levels. Mostly they failed. The main reason was that line managers had little incentive to reduce stock. Operatives, too, on the shop floor complained when work in progress was cut and they thought that they would run out of work. Perhaps, even more importantly, all the elements that needed to be put right—rejects, breakdowns, erratic lead times, excessive splash demand allied with poor forecasting—were missing. They had only a negative effect on a system under stress. The best inventory control systems in the world can never work with so many poor environmental factors. Even the advent of a punishing charge of 12% on working capital had little effect in our company.

Early in the 1980s we began a different approach.

(i) Stock levels were determined—in total, for raw materials, work-in-progress, finished goods stock for all factories, product groups, and businesses. The total figure was put into the profit plan, while the individual items became the objective of local management. They were made to care about achieving their stock targets.
(ii) Systems development specialists were then detailed to provide amended stock control systems, which line managers could use to ensure that their stock targets would be met. Stock control, then, needs to start with an objective which should get tighter each year. As a result of this approach, a system was designed which:

92 Resource utilization and cost control [Ch. 4

	This month Cash flow		Year to date Cash flow	
	Actual	Plan	Actual	Plan
Operating profit				
Royalties				
Redundancy charges				
Finance income/expense: non-group				
Finance income/expense intra-group				
Profit before taxation				
Dividends received (gross)				
Depreciation charged				
NBV fixed assets disposals				
Stock provisions				
Redundancy provisions				
Other provisions				
Amortization				
Intangible assets				
Other assets				
CASH GENERATED				

	This month Working capital			
	Actual		Plan	
	Value	Days	Value	Days
Stock provisions				
Debtors provisions				
Redundancy provisions				
Other provisions				
Provisions total				

Fig. 4.1. A working capital control sheet.

Figure 4.1 (*continued*)

Raw materials Consumable stores Work-in-progress Finished goods Merchandise			Raw materials Consumable stores Work-in-progress Finished goods Merchandise					
Stocks (gross)—trading Debtors (gross)—trading Creditors—trading Other debtors/creditors Intra-group current net			Stocks (gross) Debtors (gross)—trading Creditors—trading Other debtors/creditors Intra-group current net					
Working capital movements								
Gross capital expenditure Grants received Acquisitions and divestments Taxation A/C Taxation income/expense Dividend payment			Taxation A/C					
CASH UTILIZED			Current assets					
CASH SURPLUS/DEFICIT (1 + 3) Opening cash Closing cash Cash surplus/deficit			Net working capital					

Fig. 4.1. A working capital control sheet.

- used exponential smoothing as the forecast mechanism;
- determined standard lead times each month;
- re-categorized items each month if not sooner (GETO);
- took into account items already in stock, items being made, total order demand, etc.;
- operated on a fortnightly cycle basis;
- matched production capacities.

The systems items were then cascaded to determine:

- the potential pieces to be ordered and their work content;
- their stock value in one month
  two months
  three months.

A match was then made between the following:

- forecast stock value in one/two/three months' time and stock targets;
- production capacity available and that apparently needed.

A new 'cascade' was then made to make a match between the stock value forecast and the objective, and the production capacity available and that which was needed.

The loss or gain in service and value by achieving the stock target or matching production capacity was then calculated. It then became a line-manager decision as to what should be done:

- Exceed objectives for a while to fill available capacity.
- Live with reduced service levels for some non-key items, etc.
- Consider whether batch sizes needed to be increased or decreased and what the effect of doing this would be on production efficiency and costs.

So, under the pressure of events and especially the need to conserve cash, we developed a stock control system which improved stock service levels considerably, while at the same time reducing stock, at first by a half and then two-thirds, as a ratio of sales, of the best that had previously been achieved.

At the same time, work-in-progress was reduced by:

- using new technology
- using MRP II and better shop scheduling
- reducing rejects.

Hence, standard lead times were reduced and became more accurate. Forecasting by product group and then individual product also improved. Splash demand, however, continued to be a problem.

This is a gradualist way of reducing stocks. The alternative we thought of—setting an unachievable target for stock of 25 days—would cause much trauma. In the attempt to succeed, service levels would disintegrate, customers and markets would be lost, disaster and disillusionment would take over.

It is possible, under the system outlined, (if it is allied with MRP II) to reduce stock targets from one year to the next and still improve factory efficiency and stock service levels.

The more general approach needed to reduce stock of all kinds is shown in Fig. 4.2.

4.7.2 Stock Valuation

Over-valuing stock or at least revaluing it on some pretence is the high road to perdition and worse. No one should fall into the trap of using stock values to hide incompetence.

The toughest method of valuing stock is to base it on standard variable cost. If total costs or even total manufacturing costs are used they will add to the danger of over-valuation.

Net-realizable value (i.e., what revenue stock will generate when it is sold) should always be calculated for all product markets.

The older the stock is, the less value it should have. An age analysis is vital for this purpose—say, stocks are 3, 6, 9 and 12 months old. As each period passes, a provision for the stock should be made. Any stock over 12 months old should probably be fully provided for.

In a company with a stock turn of, say, 8 or 9, having a conservative stock value only delays the achievement of profit by a month and a half. This is a small price to pay for keeping stock values under control.

4.7.3 Materials productivity

When material cost rises to nearly 30% of revenue earned, the slightest improvement in materials productivity or rejects brings an immediate and worthwhile reward.

By far the best way of measuring materials productivity, we found, was by calculating material yield. Yield is the weight of finished products accepted by the customer, compared with the weight of material input into the factory to make the customer's order. Percent yields should be calculated for cost centres and product lines.

Material losses occur for two reasons:

- Technical —these are considered to be intrinsic to the production process—volatiles being driven off, cutting, grinding, chamfering, etc.
- Operational—losses occurring as a result of faulty workmanship by operatives or machines.

Materials productivity analysis and improvement should follow one or other of these two reasons for loss, as they need different approaches to correct them.

Technical and operational losses tended to occur unevenly in the Chapel factory, and analysis and correction were delayed because data collection was inadequate.

Technical losses occur because of:

Machine limitations (due to machine design)
Methods failure
New material quality problems
Material chemistry failure

	Target	Systems approach	General
Raw materials	Total value expressed as 1 days usage. Target by product group	Purchasing control system as part of MRP II system	Consignment stocks negotiated with suppliers Just-in-time deliveries
Work-in-progress	Target value for: Cost centres Products Product lines Total factory Businesses	Differential ordering of GETs[a] items (i.e. monthly, every 3 months, every 4 months). Scheduling systems. Input control	Minimize differentials in products until the last possible moment in manufacture. Discipline in order processing. Rejects reduced to absolute minimum. Fast die changes. Maintenance engineers associated with each production line to improve machine utilization.
Finished goods stock	Target value by: Product Product markets Businesses Total	Cycle re-ordering Cascade system. Year-by-year reduction in stock targets. Allocation system	Minimum consignment stock. Constant correction and improvement in lead times. GET categorization. Forecasting

[a] In Ferodo, the standard 'ABC' categorization was eschewed in favour of GETs—greyhounds, elephants, tortoises

Fig. 4.2. A systems approach to reducing stock.

Product design
Product geometry
Standards set too high for the production process.

Operational losses occur because of:

Operative errors
- working too quickly
- machines badly set up
- inadequate control early in the process

Machine problems
- machines going out of standard during the production process
- innate problems, due to poor maintenance
- lack of control information.

The problem of materials productivity is bound up with that of SQA and the two cannot easily be separated. SQA should tackle both technical and operational losses. It is a key element in improving material yield, but other important factors, as far as our own exercises went, were:

- *Improvement in machine controls* It is curious that some machines and ovens, though over 10 years old, still lacked adequate control mechanisms.
- *Records of input/output* Until machine controls were in place it was difficult to measure material yield accurately and allocate causes for it, between operations.
- *Operative performance* More will be said later in the way of discussion on culture, motivation and payment systems to ensure that operatives are positively motivated to improve material yield.
- *Value analysis and value engineering* An old technique, but still worth pursuing.
- *Product geometry* In many production processes, products are cut from sheets or blocks of new material. The relationship between sheet and final product dimension is a major determinant in improving material utilization.
- *Scrap recovery* Often a useful activity if it is possible to retrieve and re-work apparently scrap products and make them sellable.
- *Standardization and variety reduction* Reducing material types, product sizes, variations in finishing, will all help to improve material utilization.
- *Tool control, and engineering and machine improvements* Essential as part of introducing JIT and improvements in materials productivity.

4.7.4 Working capital generally

Having a million pounds in stock consistently is the equivalent of using the same money for fixed capital. It is an alternative. There is a need, therefore, to treat the funds for working and fixed capital as if they all came out of the same pocket.

Reducing working capital probably needs an organization where most if not all the elements of working capital are under the control of one individual.

Systems are much less effective than setting line managers objectives and measuring them on a week-by-week basis.

4.8 WASTE DISPOSAL

Being a company associated with the use of asbestos, Ferodo was always highly conscious of the need to keep the factory clean and dust counts down to an absolute minimum.

A study on mortality in the Chapel factory, stretching eventually from 1941 to 1986,[†] was carried out by outside specialists. The conclusion was:

'There was no excess of deaths from lung cancer or other asbestos-related tumours, or from chronic respiratory disease. After 1950 hygienic control was progressively improved and from 1970 levels of asbestos in air have not exceeded 0.5–1.0 f/ml. It is concluded that with good environmental control chrysotile asbestos may be used in manufacture without causing excess mortality.'

This is a very far cry from the 'Alice—a fight for life' programme put on by Yorkshire TV in 1982 and all the consequent emotions which this aroused.

Asbestos can be a highly dangerous substance, especially crocidolite asbestos; but it can be tamed. Its excellent properties for being a main constituent of friction materials are well known. In Ferodo, dust extraction and collection received investment when all other investment opportunities were eschewed. The result was, as far as possible, a clean factory.

Television rarely left out an occasion when it could stir up emotions about asbestos. For nearly two decades after the war, Ferodo dumped its waste on a tip in 'The Wash', a hamlet near to the factory. Eventually it had been covered by two feet of soil, trees were planted and it was left. In time, the combination of North Derbyshire weather, trespassers and rabbits brought some old reject clutch facings to the surface. One of these was held up to viewers looking at the evening news on Granada TV. Ferodo was pilloried as a polluter, allowing asbestos to run riot. In such circumstances, for senior managers to suffer 'trial by television' and defend the position is very difficult. I declined to appear on the programme.

No senior manager in industry wants to pollute the environment, least of all in an area like the Peak district of North Derbyshire. A clean environment is a boon for everyone.

Television's attempt to portray manufacturing industry as an irresponsible and perhaps unnecessary part of the economy is hard to refute, given the TV company's control over the programme. When we later set aside nearly £1 million to make sure the 'Wash Tip' never again caused locals sleepless nights, this was a non-event as a news item.

Although Ferodo proved (conclusively, in our minds) that chrysotile asbestos was not dangerous if dust extraction kit was appropriate, we lost the battle of its use. Other, more expensive alternatives were put into the manufacturing process.

[†] See 'A mortality study of workers manufacturing friction materials 1941–86.' M. L. Newhouse and K. R. Sullivan (of the TUC Centenary Institute of Occupational Health). London School of Hygiene and Tropical Medicine, London WC1 E7HT. Published in the *British Journal of Industrial Medicine*. 1986 **46**, 176–179.

Some of the newer residents of the Chapel-en-le-Frith district obviously resented having a factory the size of Ferodo spoiling their views. Sympathy is in order. However, the other side of the coin is important. From 1982 (when the company might have gone out of business) to 1990, employees on the Chapel site were paid over £80 million. Without this financial infusion, North Derbyshire might have become an economic desert.

However, keeping the factory as clean as it needed to be was a constant task. It says a lot about morale and culture when people do not keep their workplace clean. Bringing in the same attitudes to litter and perhaps vandalism as exist outside the factory is easy enough. Making people improve on their social norms is difficult. Yet we all knew that a clean and tidy factory was half way be being an efficient one.

The pressures we were under to produce good clean products were well known:

(1) *The Green Lobby* We were all members, to the extent that it is anathema to produce excess waste and increase pollution.
(2) *The users* Motor car assemblers too were under considerable pressure to improve their pollution image.
(3) *The law* Strict factory legislation, especially asbestos legislation, had to be followed. COSHH regulations are another important factor.
(4) *Waste disposal* Landfill sites, especially in dumping materials like waste asbestos products, were difficult to find and expensive to run.
(5) *People* People generally want to work in a clean, healthy environment. It is management's duty to provide such conditions and make sure that they are kept that way.

Everything points to material productivity improvements and if possible 'zero-defects' as being key elements in pollution control. Lorries full of waste leaving the factory should be anathema to any effective production management. Pollution control is far from cheap. Making it pay for itself by improving materials productivity is one way forward.

4.9 THEFT AND SECURITY IN RESOURCE CONTROL

(a) The problem
Resources can be misused or disappear for all sorts of reasons—technical, lack of control, inadequate operator training, genuine mistakes. All of these need to be tackled and put right, but all companies, not least those manufacturing products, have a theft and security problem.

Large department stores (and perhaps others) have coined a word for theft—'shrinkage'. The amount lost in 'stock shrinkage' tends to vary from 1% to 5%, with the latter figure being more relevant.

For any company carrying stores to a value of £5 million, or having a turnover of £60 million, 'shrinkage' can make a considerable impact on profit margins earned.

Recently, Ford Motor company found they had a 'steal to order' activity being carried out in one of their factories.

(b) What is stolen

Any motor components manufacturer is aware of the potential theft problem. Many components can be slipped into a holdall, ostensibly used to bring tea or sandwiches into the factory. Operatives know what component fits what car. If half a million a week are being made, a hundred or more can go missing without really affecting output levels.

The most dangerous theft is the stealing of substandard products which have been rejected. These can 'disappear' quite rapidly before they are destroyed or sent to the tipping area. Once a reject has been made, control over it tends to be lax or nonexistent. This makes theft easy. If the company's brand or name has been stamped somewhere (and not subsequently eliminated) a substandard product can be sold on the open market with the company's brand on it.

Small tools—hammers, paint brushes, electric drills—seem to disappear at a good rate, mainly because there are few checks on them. Once issued from the stores, they do not need to be accounted for.

(c) How

Ingenuity knows no bounds in this respect, but perhaps the most worrying aspect is not individual theft, but large-scale peculation. This can often take place from a relatively unguarded warehouse. It needs the connivance of a supervisor, a fork-lift truck driver and a lorry driver to take out a pallet of unordered goods. Taking it off the lorry at some future time may be slightly difficult.

Any vehicle leaving the factory site should be suspect, especially a noncompany vehicle.

The night shift is a time when opportunity is greatest. It is possible, on a fairly large site, to throw products over part of the perimeter fence, so that they can be picked up later.

A site where there is access from the factory to offices, and consequently to sophisticated, expensive office equipment, is providing a good opportunity for theft on a reasonable scale.

(d) What might be done?

Anyone believing that theft, with current standards of morality, can be controlled completely is very naive. A factory with a workforce of 500 has its alcoholics, drug abusers and thieves. National statistics can be used to calculate the probable numbers of each. However, the following measures can be taken.

(i) If stopping theft is difficult, at least it should be known about, when it happens. For example, accurate counting is necessary right across the factory. (MRP II and other reasonably sophisticated systems will not work, unless it is accurate.) A situation where the last count in the factory, on which operatives are paid, disagrees considerably with the count carried out in the warehouse, should never happen.

(ii) Potential theft and fraud. The possibilities of theft and also fraud need to be thought out and perhaps discussed with internal and external auditors. From experience, external auditors carrying out the traditional audit are not good at suggesting where theft and fraud could be taking place. They are constrained by financial (payment) considerations which force them to make most of their tests on stock valuation, stock checking (to see if records of stocks show the amounts actually found) and invoicing. They determine whether the accounting records are correct and if 'proper' accounting principles are followed. A separate examination should be made on potential for theft.

(iii) No car or lorry should drive into the factory site without good reason. Every vehicle leaving the site should be examined; but this, in a busy activity, is seldom practical, so spot checks should be carried out. Especially, vehicles taking out finished products should be randomly checked against advice notes. If the company has a depot or a large customer, checks at the receiving end might be appropriate. Sightings of company lorries, sitting in a lay-by, should always be followed up.

(iv) A rule should be enforced that everyone, from the managing director downwards, is liable to be searched on leaving the company premises.

(v) People who are brought on site to carry out some activity (painting, some specific maintenance building repairs) need to be carefully scrutinized, especially if the work takes place during the annual shut-down period, when local employees could be thin on the ground.

(vi) Theft should be considered a serious offence and people involved should lose their jobs.

(vii) Employees who are able to leave the site for various reasons should have closer scrutiny than those who are not.

(viii) The finished goods warehouse is, for some miscreants, potentially a bank waiting to be robbed. It should never be a thoroughfare through which anyone can walk. It should have one entrance and exit which can be closely watched.

(ix) All the technical panoply of surveillance and control should be looked at, and where necessary used.

(e) People and theft

The local trade unions normally get very upset when extra security is introduced. Their cry of 'Why can't you trust us?' has reluctantly to be met with 'I'm sorry but in this day and age trust does not prevent theft.' As everyone carefully locks their doors before going to bed or leaving the house, companies need to do likewise.

4.10 PURCHASING

From being something of a Cinderella activity in many organizations, purchasing has come to centre stage in the battle to reduce costs and resource levels. Purchasing people are now seen as being crucial in ensuring that the supply chain is effective in terms of quality and in obtaining the optimum price and service.

For example, with an increasing cost content per vehicle of bought-out items, vehicle assemblers have rightly concentrated on minimizing such costs, even, apparently, to the extent of practically driving some of their suppliers out of business.

When some car assemblers yield to high wage demands, they occasionally believe that they can regain part of the cost increase by offering extremely small price rises to component suppliers. These suppliers then have to force their own suppliers to accept low price rises and at the same time ensure productivity improvements help in keeping costs down.

It is easy to apply purchasing leverage too toughly. From 1982 to 1989, Ferodo's raw material prices rose by only 20%. The world price of some key raw materials did actually fall during this period, but most of the small increase was won by tough purchasing. Where toughness ends and unethical pressure begins is very difficult to say. Driving a good supplier out of business, because the purchaser buys most of his output at too low a price, is not very clever.

Our own approach to purchasing changed considerably during the eighties, as follows:

(a) Once it was practically axiomatic to buy as cheaply as possible. As brake and car assemblers started to demand higher and higher quality standards, so too did our own material specifications grow more demanding. Price began to reduce in importance, and quality and delivery-on-time to take its place.

(b) Relationships with suppliers began to be the mirror image of the demanding cooperation which our main customers required. We did not attempt to grade suppliers as did Ford, but we certainly expressed concern when suppliers did not meet new material specifications. We helped those suppliers who did not operate statistical process control. (Occasionally no quality control procedures were found to be in operation at all.) However, in some small companies, trying to help to institute good quality control procedures proved impossible owing to lack of staff or will. It was then a matter for debate whether such suppliers could be used in the future.

(c) We introduced tougher and tougher controls over incoming materials and components, rejecting many. The problem then occurred of actually keeping production going, as no other material was available because of stock reduction. JIT is very much a two-edged sword.

(d) At one time, all possible sources of supply were contacted from time to time; but gradually these were whittled down to those that we could trust to supply, both on time and to the quality we needed. It became increasingly difficult to change suppliers easily and quickly.

(e) Long-term contracts began to grow increasingly important.

(f) Quotations and debates about price occurred once on a major basis, then normally were not discussed in detail again. Regular 'inflation' price rises were eschewed.

(g) Supplier evaluation. All suppliers of key raw materials were evaluated at regular intervals:

- Supplier's edge on price and discounts. Price changes were monitored against rises in the RPI and world price rises/falls
- Quality
- Communications
- Help in emergencies

- Labour and other internal problems
- Invoicing
- Buying terms
- Rejects and speed of response
- Packaging
- Credit.

(h) Budgeting. Purchasing budgets were established to help maximize cash generation and minimize stock holdings. Occasionally, during times of intense crisis (as in 1982), a weekly spend was allocated, which on no account could be exceeded. However, it is not the total spend which matters, but the length of time before the supplier demands payment and when he actually gets his money. (There could be a significant difference.)

(i) Globalization and purchasing. It was customary in the Ferodo group of companies to share purchasing data, especially about prices, when buying similar raw materials. Regular purchasing managers' meetings provided the forum where this information was exchanged. At least within the EC, national boundaries in purchasing ceased to exist. If a supplier in Italy was best for price and quality, then he was supported by everyone and gained business.

(j) Quality and suppliers. Chapter 2—The product market—details the quality relationships now essential between motor component suppliers and end users. Our own supplying relationship was based on these quality aspects:

- Specification of raw material and components. These were reviewed and re-written. In some cases we found that for years a supplier's appreciation of a specification was not what we thought it should be. No wonder rejects occurred.
- Selection of suppliers. Suppliers had to conform to SQA standards if at all possible. They had to agree to on-site inspection of quality procedures, both in practice and for written records.
- Agreement of quality assurance. Data had to be offered by the supplier, including verification methods and in the settlement of quality disputes.
- Receiving and inspection procedures. Cost sampling, quarantine and deviation procedures had to be carried out.
  Quality records were needed including batch traceability

4.11 JIT—A MYTH IN ITS OWN LIFETIME

Three situations appear to be the scene for the debate about Just-in-Time (JIT).

When I was production controller in Metal Box Co. in Cape Town, we operated a delivery system where customers collected open-top cans, not just on a particular day, but at a specific time. The reason was obvious. A day's production filled the available storage space. Whether this, as the proponents of JIT would claim, ever helped to make Metal Box into a world-class manufacturer is debatable.

The Chairman of our Japanese associate company visited the Chapel factory and was shown the new investment we had put into robotic cell manufacture. He admired the technology, but frowned at the amount of work-in-progress surrounding the cells.

'You have some way to go yet,' he said indulgently. Yet the return on investment achieved by the cells, and the way they were operated, was way ahead of the British national average.

The Chairman of one of our customers was quoted as admiring the best Japanese companies and their stock turns of 100 or better. Yet the schedules from that same company occasionally bounced around like manic yo-yos, preventing us from reducing work-in-process.

These aspects of JIT and stock holding generally made us somewhat cynical about claims that even British companies can operate with stock of 25 days or less. JIT, we believed, assumes:

(a) Nothing ever goes wrong or will go wrong in the production process, e.g. breakdowns, absenteeism, rejects, industrial relations.

   Stopping the line of a brake or car assembler would provoke wrath of immense magnitude. The Ford engine plant in Bridgend did, and the result was a major reduction in future capital investment.

(b) Infinite flexibility. In the case of batch sizes, for example, some consultants would say that 'one' is the best.

   No matter how hard we tried to reduce die change times, they were still tedious and time-consuming. We introduced training and re-training: dies were kept immediately by the cells; operators were pressured to change over as fast as possible. Yet, still, change took a half hour or more. Making 'one' a batch size seemed ludicrous and appallingly expensive. Flexibility cannot be infinite and it can be extremely costly.

(c) There is little or no cost penalty in not having work-in-progress or finished goods stock. Or, if there is a penalty, then it is far less than the quantifiable benefits.

   Reducing work-in-progress by one million pounds at 15% interest rates saves £150 000 per year. It is highly likely, in our view, that this amount could be quickly consumed in making emergency deliveries, due to erratic customer demand.

(d) The relationship required between supplier and customer is one of mutual agreement on deliveries/scheduling and of support of each other's production line.

   This kind of symbiotic relationship seems very difficult to achieve. We offered one of our key customers the facility to interrogate our MRP II files. The two companies' systems, we thought, might be linked, as we were using the same standard software. The benefits would be an immense improvement in scheduling relationship. The offer proved to be too political to implement. Instead we sent all our major OE customers print-outs from our MRP II system showing the schedule position—orders, work-in-process, despatches, outstanding items. We could do little more to improve customer relationships.

(e) Major changes in training, culture and organization have been put in place and are effective.

In our case, these factors were indeed improving, but they were being introduced, not as part of a JIT activity, but as a component of an overall manufacturing strategy.

(f) MRP II, or at least, MRP, has to be installed and used effectively.

With this we agree.

(g) Stock reduction is regarded by the Sales/Marketing function as a major advantage in servicing customers.

In Ferodo, all the moves towards reducing stock and work-in-progress came from either the Finance or Operational Planning functions. Sales People were vociferous in maintaining that stock of a wide product range was essential to underwrite their selling effort.

(h) All activities in the business have to be smart. This applies as much to despatching as to production and order processing.

Gradually, we moved as many as possible of the traditional warehouse activities such as packing to the production line, to enhance speed of order processing. Even so, the despatch manager's question is still relevant: 'Do I send half a load now or wait and send a full load tomorrow? If you want to send half a load now, I need more lorries.'

(i) Customers and raw material suppliers need to be relatively close geographically as well as philosophically.

Selling to South Korea or buying some raw materials from Australia does make difficulties which despatch to a customer down the road would not.

(j) Managers, supervisors and operatives are in sufficient numbers and well-trained enough to overcome any crisis which occurs in the order processing, production or despatching activities.

People need to be paid to operate a system which has no stocks, with all the likely consequences this could bring.

(k) Production people need to have infinite flexibility in terms of operating machines and processes whenever demand for them occurs.

Factory culture to do this is probably more important than using a computer to schedule the factory.

(l) Factory layout should be such that it enhances the chance of operating JIT successfully.

Group technology in Ferodo should have put the company into a good position to enforce JIT many years ago. This was not so mainly because factory culture was not changed at the same time that the layout of the factory was re-established.

(m) Local operatives need to take to JIT enthusiastically.

JIT is a philosophy which has company-wide ramifications. Properly applied, it should drive an organization into carrying out all kinds of improvements—work organization, payment systems, planning/scheduling and so on. This seems an odd way of creating a manufacturing strategy. Making JIT the motivating factor is a shuttle decision.†

Of course, no one would dispute the need to reduce stocks of all kinds to a minimum. Ferodo's stock turn improved significantly during the eighties. This came about by applying well-known production engineering techniques, better systems and improved financial disciplines, including:

- Reducing the product range
- Improving maintenance
- Changing work organization
- Applying capital investment to new manufacturing technology
- Better forecasting
- Much improved MRP II/MRP and associated control systems.

It seems much safer to penetrate an overall manufacturing strategy with studied cost/benefit analysis and let stocks reduce as a result, than to say we will introduce JIT and let it drive the actions for stock reduction.

4.12 RESEARCH AND DEVELOPMENT

Ferodo was inordinately proud of its R&D and especially its material testing facility. These, as all the publicity blurbs stated, were the largest in the world.

Despite this, R&D managers were constantly critical of the small amount of money that was put into R&D. In terms of percentage of revenue earned, West German friction material manufacturers spent twice as much as Ferodo did.

Here then was a classic resource allocation problem. If the company in total was making too little profit, from where would any extra money for R&D come? What else would suffer as a result? If more money was allocated to the function, what certainty was there that it would be used wisely and to the best effect? Should Ferodo, as a rule of thumb, spend the same, proportionately, as West German companies?

The R&D activities were central to providing Ferodo companies round the world with new product formulations, indeed new products and production facility improvements.

The R&D function had, in some respects, exhibited some of the 'British disease', developing apparently good products without determining how they could be exploited. The 'retarder' was a case in point. This was a product designed to help braking on coaches and heavy trucks and eschewed the use of friction materials in favour of hydraulic tension. A separate factory was rented for its manufacture. A minister opened the factory. Eventually it was closed as it was not profitable.

---

† Resulting from Directors reading business magazines while travelling on the Manchester–London shuttle service.

What went wrong? The idea was good. The engineering was sound. The failure lay in several areas. Insufficient money had been set aside to cover marketing and selling activities. Only a vague assessment of the market appeared to have been done. Insufficient selling effort was made. Enough resources had been allocated for development, but too few for its exploitation. The people involved in development were not those who had to gain revenue from it.

So, what lessons did we learn?

(1) R&D departments should not be allowed to build themselves ivory towers where scientists live in large laboratories, full of expensive equipment and largely divorced from the real world of production and selling

(2) The function needed to carry out or become closely involved with technical activities other than those which involved R&D.

Quality assurance was a case in point. With increasing sophistication in both materials and production processes, production management needed assistance in handling their situation. Not only Quality people were needed on the shop floor, but material technologists as well. Putting Quality Assurance with R&D helped this process.

(3) R&D people should be positively involved in selling their product developments to their customers. Increasingly R&D technologists visited original equipment users and debated both product needs and developments with local specialists.

This necessitated a new organization structure where managers responsible for part of the R&D funding were also active in specific and related businesses. They helped to gain revenue and to improve cost performance.

(4) Nothing should be developed beyond modest investment unless a market or a potential market had been determined for it.

As soon as possible a calculation of the potential contribution earned and the costs incurred had to be made.

(5) Largely R&D strategy had to be active in supporting world car scenarios and determined attempts to gain business from large and successful motor vehicle assemblers. These would guarantee longer-term revenue, if achieved. Where potential contracts become bigger and more important financially, it was essential that all technical selling resources should be thrown into the battle.

(6) How then were projects and activities to be chosen? Focused factories and businesses should bid for help on the basis of cost to be incurred against potential revenue to be earned.

Taking this view tended to dictate the kind of overall marketing strategy to be followed. If one particular product market was achieving a higher contribution per piece sold or gained from individual customers, then this was given support. It is possible that this would be a wrong decision—the product market in question might be at the height of its product/economic life-cycle—and that a maturing market might have been a better proposition. Perhaps a too-literal policy was followed, where items were listed in potential contribution order and cut-off point reached where costs were no longer justified.

(7) How was the R&D budget in total to be set? What resources should be allocated to the function?

Any function which demands resources which can only be obtained at the expense of some other company activity needs to express its needs very carefully. The kind of angry exchanges in the Boardroom where the Works Director suggests that 'R&D wants me to fire more planning clerks to pay for material technologists who will not earn their cost' should be avoided. R&D managers should make their case as follows:

(a) The opportunities which will be missed if the R&D budget is not extended should be listed carefully.
(b) The control over what happens in the R&D department should be explicit. Any view that the scientists are a race apart, governed by no control mechanisms, should be crushed. Control has to be even tighter than in other parts of the organization.
(c) Reports on progress on all key activities should be made available to all senior managers on a monthly basis. The belief that R&D squanders hard-earned revenue should be seen to be untrue.
(d) R&D personnel should be seen to be active, not just in developing new products and materials, but in using their technical knowledge in areas where it may be missing, such as:
   - quality assurance;
   - process development;
   - in setting up a material value analysis activity where R&D people play a key role in reducing the cost of raw materials;
   - helping the purchase department in discussions with raw material suppliers, about quality.

R&D managers should attend, and be major contributors to, profit plan review meetings.

For many years Ferodo had believed that getting product formulations which were better than those of competitors was something of a black art; that in carrying out the process, R&D people needed to be left alone.

It was significant that this earlier approach had produced far fewer new products than those which were achieved later with a much reduced budget and a tough project choice and control system. No matter how esoteric the activity or the scientists concerned, hard financial evaluation on cost and benefit was essential. Scientists, just as much as production or business managers, needed to understand 'Business'.

4.13 PRODUCT DESIGN AND R&D

Product design tends to reflect the technical competence of the organization. If a product looks well and exudes an air of reliability, it tends to attract customers.

(a) Design should be reflected in both performance and overall quality. Many products can sell on their design alone—witness many household goods like washing machines, cookers and food mixers.

(b) Product design should reflect back into the industrial/manufacturing process. How a well-designed product is made should form part of the technology survey. The production process might be eased by good design at the same time as providing a sound manufacturing activity.
(c) Design should form a major part of the strategic fit analysis; e.g., what design will best serve the market? What customers might be looking for is reliability, value for money, after-sales service and good design.
(d) Design should cover products, but also the merchandising, packaging, advertising and the sales campaign generally. Design might offer an opportunity to project a good company image.
(e) CAD/CAM resources might be deployed to improve product design.
(f) Product design is often a 'one-off' activity and likely to be best done by bringing in an outside specialist. Even so, design should be completely integrated with the general R&D function.

4.14 RESOURCES DATA AND CONTROL

The following ratios will help in determining how well resources are being used if data can be established for the last 3–4 years (the trend) and for businesses, focused factories, product lines and product/markets.

People

$$\frac{\text{Added value/revenue/contribution}}{\text{Wages and salaries}}$$

$$\frac{\text{Net sales}}{\text{People employed}}$$

$$\frac{\text{Direct personnel}}{\text{Indirect personnel}}$$

$$\frac{\text{Total personnel}}{\text{All departments/sectors/functional personnel}}$$

Cash

$$\frac{\text{Cash inflow}}{\text{Cash outflow}}$$

$$\frac{\text{Cash flow}}{\text{Short-term borrowings}}$$

Administration

$$\frac{\text{Added value/revenue/contribution}}{\text{Administration costs}}$$

$$\frac{\text{Administration costs—by department/function}}{\text{Net sales revenue}}$$

Fixed factory costs

$$\frac{\text{Maintenance}}{\text{Direct labour costs}}$$

Working capital

$$\frac{\text{Net sales revenue}}{\text{Net working capital}} \qquad \frac{\text{Cost of scrap}}{\text{Total material cost}}$$

$$\frac{\text{Debtors outstanding}}{\text{Net sales revenue}} \qquad \frac{\text{Sales revenue}}{\text{Finished goods, WIP, Raw materials}}$$

$$\frac{\text{Material cost}}{\text{Net sales}} \qquad \frac{\text{Total stock}}{\text{Total working capital}}$$

Material yield

Resources generally

$$\frac{\text{Total assets}}{\text{Operating profit}}$$

$$\frac{\text{Production capacity utilized}}{\text{Total capacity}}$$

$$\frac{\text{Power costs}}{\text{Machine running-time}}$$

CONTENTION

(1) Conserving resources and reducing costs is as much a cultural problem as it is a planning and cost control one. No cost should be incurred if it does not have some positive impact on cost reduction or revenue/contribution generation.
(2) The chief executive/managing director has probably most influence over whether minimum cost is incurred in achieving profit or whether costs flow out in a never-ending stream. The example the CE/MD sets should permeate the whole organization.
(3) People generate costs way beyond their wages/salaries. While budgetary control, top-down planning and cost control will keep a lid on costs, they will only be reduced sharply by reducing people.
(4) Computers have been something of a confidence trick as far as their claims in reducing people of any sort; and so have Management Services Departments. Numbers of administrative people can best be reduced by the application of old-fashioned work simplification—allied with modern office technology.

(5) Techniques like manpower and succession planning have not proved very fruitful in keeping people and costs under control and providing requisite skills.
(6) Cash is probably more important than profit. Cash control and generation is extremely important.
(7) Stock control systems have proved to be of only minor importance in reducing stock values. Establishing stock targets for factories, businesses and product markets and motivating line managers to demand systems of control which help is much more fruitful.
(8) The culture of most companies is probably against the successful introduction of a rigorous just-in-time system. The preferred method is probably a constant reduction in gross stock value objectives, reducing, say, from 80 days to around 40 or less.

A suitable set of strategies

(1) People being the most important element in cost, we will set an objective of improving our labour/revenue ratio by 40% in five years. We will do this by continuing functional analysis, work simplification and general re-organization especially of staff functions. Productivity improvements of the same order are needed in direct and indirect activities. We will achieve these by changing incentive schemes, applying capital investment when it is proved economic, and buying in services where these are cheaper than we can currently provide internally, while still giving the same service.
(2) We will continue to reduce people on a year-by-year basis, taking account of cost reduction needs, demographic trends, skills acquisition and age/sex composition of the workforce.
(3) Cash generation will be a major objective for everyone in the organization. Positive cash flow per month is a key aim, although per quarter may be more realistic.
(4) We will continue to explore the possibilities of buying out or 'privatizing' many of the services which we currently carry out internally. It may not be necessary actually to use outside/non-company agencies, if local people will bid for the service.
(5) Working capital will be contained at a ratio of 15% of net sales revenue. Stock targets will be gradually decreased until we achieve a 40 day stock turn. The Cascade system will be improved to help in this process. MRP II will be used increasingly to reduce work-in-process.
(6) Improving material productivity will continue to have a high priority among production engineering personnel. We hope to improve yield 5% this year.
(7) We aim to ensure that the Purchasing function and suppliers become the mirror-image of the relationship we have needed to build up with our OE customers.

5

Work organization and training

5.1 INTRODUCTION

In the *Production Management Handbook*[†] I suggested that an effective work organization was one where power, authority and responsibility had been reconciled. *Power* was defined as the means of making others comply with one's wishes whether they wanted to do so or not. *Authority* was the ability given to someone to commit resources to achieve a particular objective.

It was also suggested that while managers might be given the responsibility to achieve objectives, they may have neither the requisite power nor the complete authority to do so.

This is a dilemma which is at the heart of establishing effective work organization.

To the three elements quoted—power, authority and responsibility—might be added a fourth: the privileges associated with a job or status. Privilege has a deleterious effect on work organization, when it is perceived to be gained without commensurate responsibility and, above all, achievement.

I have yet to see a British Managing Director standing in the same lunch queue as his operatives, which is a standard practice in Japanese companies (but no doubt some now do). The curse of the company car is a very potent factor in spreading dissent and envy. The ideal where managers have no perks at all still seems a millennium away.

In the *Independent* of 5 January 1990 striking pickets standing outside the British Aerospace factory at Chester were quoted as follows, when asked about attempts to harmonize conditions:

[†] B. H. Walley. *Production Management Handbook*. Gower. Second edition 1986.

'About the only thing that has changed in eight years is that they have put a drinks machine on the shop floor.'

'A lot of people have been made more aware of bosses going out at 12.10 pm and coming back at 3 pm. We are docked pay if we are a minute late. They say they have no money for the shorter working week but then another truck loaded with brand new Rover company cars turns up.'

There may be every legitimate reason why senior management take long lunch breaks, but it is the perception of management and their perks, when set against their apparent power, responsibility and above all, achievements, that really counts.

If British industry fails in its phoenix role in the nineties, there will be a variety of reasons, but none, probably, as important as the still-deep divides between managers and their workforce and among the workforce itself. The class and educational systems have had an appalling influence on the efficiency of many British manufacturing units. Unless managers can be seen to take the moral high ground, in asking for change and all that goes with it, nothing much will happen except decline.

Like many other companies in the UK we started to think about team working and harmonization of conditions to ensure that we became a manufacturing phoenix. Like many of those companies, there were stumbling blocks to major and speedy progress, as enumerated below.

(a) Unclear aims
Most if not all companies set themselves objectives and ensure that management and worker alike have a distillation of company requirements. In the case of a manager, the relationship between company aims and his own objectives may be clear-cut. That between a shop floor operative and the company could be totally obscure. This of course assumes that senior management has worked out and recorded company aims, which are clear-cut, unambiguous and understandable, if not completely accepted, by everyone.

(b) Unclear values
Profit and profit earning is probably the most debated and debatable element as a value system. What will be sacrificed to achieve satisfactory profits—people and jobs, the local environment, pride and self-respect? Are good working relationships better than the last thousand pounds of profit? Culture might be another word for this factor.

(c) Lack of management skill
It is possible that some highly intelligent managers will be able to manage well without undue training, but most will not. As running a business grows more and more complicated, so too must training be advanced. It is unlikely, for example, that MRP II can be introduced without a major training initiative.

(d) Inappropriate management philosophy
Again this impinges on the culture of the organization. Management may not even have a conscious philosophy and what occurs is more by default than design.

Some philosophies—paternalistic, autocratic, class-ridden—may appear to be completely inappropriate for the nineties, but some managers may be difficult to change. Historical precedent may linger on, despite good intentions.

(e) A lack of cohesion
Despite considerable lip service, many companies do not operate as a team. The sales people fight production; the accountants stay primly above the battle; R&D personnel luxuriate in their specialisms.

(f) Confused organization structure
Allied with this, can be a poor design of jobs, where responsibilities are crossed with others.

(g) Leadership
At a time when teams, and all that these can provide in the way of personal fulfilment, have so much going for them, it may seem strange to put leadership down as a key factor in the winning and losing of the business battle. Genetic endowment will always give some better leadership qualities than others. A company needs these people to take risks, to take decisions, that teams cannot take, to introduce initiatives which are badly needed in the company. Some role models are vitally needed.

(h) Inadequate planning and control
Again, despite the many benefits which teams bring, planning and control may not be fully exploited.

(i) Inadequate rewards and punishments
This, combined with motivation, can often wreck the most well-conceived of organizations. Individuals and teams need to be well aware of what is expected of them and that if they succeed or fail, rewards and punishments will be appropriate.

(j) Poor motivation
Despite good training and adequate organization, even with well-stated aims, motivation may be lacking. Money rewards are important, but so are hygiene factors, and while these do not provoke the debate of some years ago, they exist and should be taken into account.

(k) Conflict
Internal conflict is a direct failure to produce a satisfactory work organization with appropriate rewards and punishments. There are several definitions of conflict—antagonisms or even indifference towards objectives, or a failure to reconcile opposing personal and company objectives. These suggest that conflict may occur anywhere at any time, and to some degree is going on all the time, even among senior management.

Conflict may not necessarily be physical (a strike, an overtime ban or a direct refusal to co-operate), but is any action or indeed non-action which debilitates the organization, which is then less likely to achieve its goals.

5.2 SOME ORGANIZATIONAL CUL–DE–SACS AND RESPONSES

(a) Hierarchies
These, it has long been felt, were the way through which an organization should be planned and controlled. All authority and power, it was assumed, were contained within the hierarchy.

As people were promoted through the ranks, their power and authority apparently increased.

In practice, hierarchies have proved to be inflexible. They generate too many layers of management. They allow those lower down the hierarchy to shun making the decisions which their role might demand of them. Decision-making is pushed upwards.

The possibility of reconciling power, authority and responsibility diminishes, the more rigid the hierarchy becomes. Perhaps the British have been reluctant to abandon hierarchies, because they reflect much of the structuring in society generally, especially class.

(b) Line and staff
This other favourite of writers on organization, when combined with hierarchies, can generate considerable bureaucracy and so cost. Three examples should suffice.

Most manufacturing companies have had a Personnel function which certainly in the seventies appeared to grow enormously.

Nothing was too small or unimportant to be taken over—wage bargaining, industrial relations, management development and training, health and safety, welfare, canteen—all with their sub-units and committees. In this process, line managers were undermined and their role diluted. They found that they could not even recruit their own people, nor discipline them effectively without the involvement of 'Personnel'. Shop stewards argued with first-line supervision only under the aegis of the Personnel function, not out on the shop floor, where it should have happened.

Little wonder that line managers shrugged their shoulders and handed over all their I. R. problems to the Personnel function. When they did not achieve production targets, they could always reply, 'We had an I. R. problem with which Personnel was dealing'.

Any forward-looking company should largely abandon the Personnel function as it has grown up in the UK. Line managers should solve their own personnel problems and carry out all their other functions, not hand them over to a third party.

Management Accounting, too, has spread its tentacles, but it has not always provided suitable information to manage the production activity on a day-to-day basis. Instead, it has built up its own bureaucracy, trading information from desk to desk in its own department, as if the outside world did not matter. Management Accounting should become part of the production function, not be separate from it.

Perhaps the least loyal of the burgeoning functions accepted in the past in British industry has been 'Management Services'. To a reasonably cynical outsider, Management Services departments often appeared to be staffed by mercenaries, whose main task in life was to scan the job advertisement pages. The situation changed somewhat when a particular specialization appeared to become obsolete. Then there were no better Luddites than those in computers, management services generally, work study, O&M, or wherever the threat was directed.

A company should not need to put its faith in personnel who are likely to leave for the organization up the road if it offers another £500 per year more in salary.

(c) Focus sharpening
Faced with deteriorating performance or even just a vague belief that organizational change of some kind is needed senior management have responded by introducing focused organisations. For example, most motor component companies operate in four distinct markets—home, export, OE and replacement. These can be sub-divided further into product market segments.

It is a logical step to treat each market, perhaps each major product-market segment, as a separate business, with all that this entails in profit planning, cost containment and revenue achievement. That the factory cannot be organized in the same way is the Achilles heel of this proposal, and unless a materials management or operational planning function can act as a bridge between the business and production, the whole concept might fizzle out in acrimonious debate about production allocation and priorities. If 'Business Managers' are just salesmen in disguise then they will only lessen the effectiveness of the idea even further.

Quality circles/problem-solving task committees
'Quality circles' has already been mentioned in Chapter 2. Ferodo tried the idea but it only had minimal success. Why?

It is too easy for a cynical and disillusioned workforce to see quality circles, or any other form of joint problem-solving, as just another means of exploiting the workers. If the right culture is not being established through a whole variety of activities, from improved communications to single status, then quality circles will fail just like any other mechanism which ostensibly could be beneficial.

First-line managers, who may already feel threatened by powerful show stewards and their relationship with the Personnel Manager, will not welcome problem-solving committees which take even more of their line authority away. This is especially true when operatives can swop their mundane boring routines of actually producing things, for tea and bright discussion, leaving the first-line manager to attempt to hit his production target with fewer operatives than he needs.

(e) Organization and manufacturing strategy
It is difficult to divorce organization changes from the overall manufacturing strategy. The pressures for organizational change are similar in many ways to those in the company generally:

— technological factors: increases in technology need an increasing skill base. Specialists need to be accommodated within the organization.
— social and economic factors: the social mores, especially in discipline and attitudes to work.
— market factors: product/market changes will change focus and direction.

- general manpower problems: demographic trends, scarcity of specialist staff.
- company culture: need to accommodate low-level decision-making.
- profit planning: management orientation towards role and objectives.

5.3 GOAL CONFLICT AND WORK ORGANIZATION

If Karl Marx is right on at least one point, then workers in a factory suffer from alienation of some kind. They regard work as a means of earning enough to have a house, maybe on a mortgage, and the means to go to the pub or club most evenings, or to use the same money to pursue a hobby.

Various academic industrial psychologists have carried out surveys which go a long way towards supporting Marx's contention. In the UK it has seemed inevitable that production managers will have to put up with a workforce that will buck the system if it can and spend as much time as possible in the smoking cabin reading *The Sun*. Yet somehow the Japanese appear to have been able to create organizations where there is commonality in objectives.

In the UK conflict can take many forms, not least indifference to company objectives. While the broad mass of the workforce and management have diametrically opposed goals, then conflict is inevitable. Equally, while there is relative deprivation—that is, some groups of people believe that they are being treated less well than others and consider this to be unfair and unreasonable—conflict will ensue.

Work organization should be a powerful means of creating shared goals or, at least, goals which in total can ultimately become shared. (As a concomittant of good work organization, harmonization of conditions will solve relative deprivation.)

The most testing and important conflicts in Ferodo in the 1980s were between management and workers in the CSEU.

In 1982, senior management was faced with making immediate and savage reductions in cost. Considerable numbers of staff people were made redundant. The largely non-staff people who were still needed to make things and service machinery had pay cuts.

To their great credit, members of the T&GW Union agreed to reductions in pay and signed an appropriate document. The senior manager negotiating the deal believed that the CSEU had also agreed with a pay cut, but they had asked to be excused from making a written undertaking. Grateful to get any satisfactory deal, the management negotiating team did not pursue the matter further.

About 18 months later, after the company was again starting to make reasonable profits, a member of the CSEU instigated proceedings to retrieve the pay cut and demanded back pay. Here, according to the senior management who had been concerned with the negotiations, was a situation where a union reneged on a solemnly undertaken deal—even though it was only a verbal one. (The law was to conclude otherwise.)

The case was fought all the way to the House of Lords where Ferodo lost. It can be read in the relevant legal proceedings under the title 'Rigby v. Ferodo'. Chastened and angry, management had little good to say about the CSEU. Were they right?

(a) Not totally. Little had ever been done with the CSEU to generate common goals. No work organizations had been set up which would help in this respect. Certainly relative deprivation seemed high.

A monolithic 'Engineering Department' which has little firm organizational affinity with production and other functions is obviously a mistake. It can only foster its own internal objectives and grievances. (Subsequently CSEU members were posted permanently to the factory, where they formed joint teams with T&GWU and MSF personnel.)

(b) Even with disaster at the door, it is never easy for any group of people to accept substantial sacrifices when they have been given little indication of impending doom, and certainly no procedures for commenting on it. If a work organization with appropriate objectives is missing, then knowledge which will help to condition acceptance of some sacrifice will also be missing

(c) There had been little discussion with the CSEU about their perceived relative deprivation *vis à vis* the T&GWU direct operatives. A typical situation had grown up where fairly loose PBR systems gave to direct operatives with limited skills quite high earnings which skilled CSEU people could not equal. Work organization as well as a different payment system should have helped.

(d) Any agreement, no matter how solemnly agreed verbally needs to be recorded and signed, especially when there is no history of trust between the partners. The various legal luminaries were amazed that such injudicious activities were carried out. It shows that they have little experience of trying to run a British manufacturing company.

Even between managers and the company, commonality in objectives can be missing. How much more so it can be between the company and operatives. To take the debate a little further, the table below sets out a company's and a manager's viewpoints. The dichotomy is easily seen.

Company	Manager
• Training to improve profitability	• Training to improve job satisfaction, security and promotion prospects
• More discipline	• Less discipline
• Personality of managers to be subordinate to company strategy	• Opportunity to express personality, even if this clashes with company strategy
• Goal-motivated	• Ego-motivated
• Oligarchic/hierarchic organization with limited participation	• Democratic organization with maximum participation
• Best possible strategy to be adopted irrespective of who designed it so long as	• Need to influence strategy irrespective of whether a best result comes out or not. To be

corporate objectives are achieved	informed on all aspects of company strategy
• Role of management to be well defined and objectives issued to managers	• To set own role and objectives within agreed company strategy
• Machine-orientated, with fewer human resources around	• Not to do work which is repetitive or demeaning in some way
	• More human resources—especially subordinates
• Change to be introduced as rapidly as necessary to meet competitive and general environmental pressures	• Change to be introduced only as quickly as understanding and learning can take place, without a great deal of effort
• Control on the plan to be tough and rewards issued according to success	• Some relaxation of hard control allowing for human frailty and and things not under the direct control of the manager
• To consider profit earning the sole reason for company operation	• To do work which could have some social value as well as be profit earning

Unlike many publicly funded activities like, say, local government, where, in the past, the organization has seemed to be run for the benefit of some members of its organization, manufacturing companies are normally oriented towards making profit. In the process, some managers may see that their personal objectives are strongly opposed to those of the company. Reconciliation is needed, but not at the expense of subverting all the individual personality of the manager.

Most managers will see that a complete identification with company goals might be contrary to their own career prospects.

5.4 EFFECTIVE ORGANIZATION

What then, from experience, is likely to be an effective work organization?

(a) Flexibility is a necessary feature. Environmental pressures of all kinds are increasing; organizations need to be more flexible and less structured to cope.
(b) Power, authority and responsibility need to be reconciled.
(c) Industrial pluralism needs to be recognized and built into a revised work organization.
(d) Conflict, especially by goal setting and elimination of relative deprivation, needs to be minimized (it will never be eliminated).
(e) Hierarchies do not work and as far as possible they should be abandoned.
(f) Job descriptions which put a firm edge round every job are not essential.
(g) There is no such thing as similar skills in individuals, managers, groups and teams. They should all be treated differently.

(h) Organizations should directly unite what few staff people remain with line personnel.
(i) Organization charts, with neat boxes and lines of command, are not relevant. Organizations need to be irregular and askew in some, perhaps in all, parts.

All these factors tend to point towards establishing teams of people, perhaps within an overall team, which might be the total organization. Within this organization, harmonization of conditions will have been established and as far as possible relative deprivation ended.

5.5 TEAMS/GROUPS

It is probable that in designing a new and more relevant work organization, management (like we did) will spend most time on trying to establish effective teams.

Teams of some sort will have, nearly inevitably, been established already. It will be an initial task to consider these and how effective they are and whether or not they need to be broken up and re-constituted in a different way. At least their objectives and the skills they need to achieve them will have to be reconsidered.

People join groups or want to belong to groups for a variety of reasons—security, dependency, reinforcement of prejudice, self-identity.

To offset some of the euphoria which industrial psychologists sometimes show about teams and manufacturing, quoting some drawbacks might be worthwhile. We found a considerable number of drawbacks, not least these:

(a) Groups or teams are not always good at solving problems or making reasonable progress in generating new ideas. They will often lack the skills needed to provide solutions to manufacturing problems within their orbit (the same lack of training often besets quality circles). Experience of operating a machine for ten years or so is no substitute for understanding the engineering design of the machine and how it can be improved. It is largely a myth that 'the man on the job' knows best.
(b) Contrary to academic opinion, operatives do not always welcome flexibility. Moving around and carrying out several different jobs can dilute skills (especially manual dexterity) and so reduce earnings. Without some prodding, operatives within a group tend to gravitate towards one or, at the most, two jobs where they have mastered the requisite manual dexterity and general skills.
(c) Groups of teams can become insular and inward-thinking depending upon their internal knowledge and basic coherence. So if 'too much' change is introduced, the group identity can change and so the effectiveness of the group in total.
(d) In terms of pay and productivity, teams and team working can actually be counter-productive. One or two strong members of a team can easily dominate the team as a whole, determining what output is achieved, who does what, and whether productivity is improved or not. The rest of the team will go along with opinion-formers and unofficial leaders. Group-think can suppress dissent.
(e) Personal doubt or individual desire is often suppressed in order to achieve, or perhaps maintain, group loyalty.

Equally, there could be major benefits in having an effective team structure. Amongst the most important could be:

(a) There are many cases now to prove the point that setting up effective teams and generating team enthusiasm for profitable change is often more beneficial than introducing new technology.
(b) Correctly established and trained, teams can exploit, or even use for the first time, their knowledge, flexibility, enthusiasm, skill, commitment, which no other work organization can do.
(c) 'Just-in-time' stock philosophy, total quality management, zero-defects, etc., all need an effective work organization, which must largely be built around teams.
(d) Shared objectives, a more congenial working atmosphere, the introduction of a team spirit will have definite spin-offs in reduced local absenteeism, fewer disciplinary cases and perhaps a major reduction in possible conflict.

So, if despite the drawbacks, team working is so preferable to other work organization, why is it not instituted in every manufacturing organization? Mainly, it seems, because:

— An appropriate company culture is not place.
— Communications are not good enough.
— Team leadership has not been considered in detail and potential leaders are not well enough trained.
— A training programme to build up requisite skills has not been attempted.
— Existing pay structures inhibit change.
— The idea had not been sold properly. It is misunderstood.

UK manufacturing companies still appear to change slowly. The innate conservatism of managers and workforce is a constant challenge to anyone who wants to make significant change, especially within the work organization. Any attempt to introduce effective team working must take time, has to be well thought out, and must be accepted at all levels of management. The conditioning or softening up must be well conceived.

Prior to introducing new work organization we thought we needed to do the following:

(1) Show a commitment of the most senior management to changes in culture.
(2) Improve communications continuously, with team briefings being undertaken frequently.
(3) Discuss (what now seems endlessly) new payment structures.
(4) Start to harmonize pay and employment conditions.

One major change we needed on the shop floor was in the training and motivation of first-line supervision. (If much of the remainder of this section is concerned with this topic, it is because it is at this organizational level that team working in a factory will succeed or fail. There is no alternative to bringing about a major improvement in the skills and morale of first-line supervision.)

Team building is taken further in section 5.6.3.6.

5.6 FIRST-LINE SUPERVISION

5.6.1 Introduction

In Ferodo's Chapel factory we had about 40 first-line (or section) managers in the early eighties. From their impact on output and efficiency there might have been only half that number, at the most. They had little respect. Why?

(a) The Personnel Director had appropriated many of the functions of first-line supervision, especially 'industrial relations'. Shop stewards had come to believe that only 'Personnel' could solve their problems.
(b) Perhaps as a result of the dominance of the Personnel function, the section managers had largely given up trying to discipline their operatives.
(c) Senior management did not lay down firm unequivocal objectives for the section managers. No one monitored their performance on a day-to-day basis.
(d) Training—once this had been very good, but by the early eighties had largely ceased.
(e) By European, especially West German, standards, the section managers were not very well educated. They were largely drawn from the shop floor and were being asked to plan and control people who probably lived in the same road as they did. Some were obviously under strong social pressure not to rock the boat, no matter how bad things were.
(f) The planning/scheduling system was so poor that the section managers spent large parts of their day chasing delayed orders.
(g) The role of the section manager in running a modern factory was not delineated.
(h) Their pay was modest by comparison with some of the high-earning operatives, able to exploit not very well conceived payment-by-results schemes. Section managers were paid for overtime, so it was understandable if some overtime appeared contrived.
(i) As technical improvements were made, their lack of technical skills became more and more manifest
(j) Perhaps because of their ambiguous and unsatisfactory position, they had become vociferous members of the ASTMS (now MSF) Union. Again perhaps because of their unionization, senior management no longer regarded them as being in the management structure. How could they be, when they fought for pay and conditions improvements with all the vigour and cunning of the TGWU and the CSEU?
(k) Between the Production and Engineering section managers there were quite significant differences in style and results. The engineering foreman had traditionally enjoyed considerable autonomy. He was 'skilled' in that he had served an engineering apprenticeship. He was probably, on average, slightly more intelligent than his production counterpart, but no better at controlling his workforce and ensuring that they did an effective day's work.

This, by any count, is a formidable list of negatives. Add to it the view that the factory was largely run, perhaps dominated, by Taylor-type payment-by-results schemes. They did indeed work. Production was achieved, but at a cost, and in the long run these schemes provided no basis for needed major improvements.

It was perhaps fair to say that section managers allowed PBR to do their jobs for them. This was not an untypical scene in many UK factories at the time. How to make significant change?

5.6.2 Objectives and performance measurement for first-line supervision

A speedy improvement of the unpromising situation recorded in the previous section was needed. There was little chance of introducing payment schemes, team organizations or even the first vague glimmerings of a different culture, unless the jobs of first-line supervisors were re-defined, key objectives set, and performance monitored consistently and well. That cardinal discipline in management, or allowing no one managerial authority without being set agreed targets which were to be carefully monitored, was followed to the letter. Simple objectives were re-issued as follows:

(a) Achieving agreed levels of output.
(b) Minimizing costs by keeping labour hours, as a ratio of output, as low as possible and minimizing rejects.
(c) Ensuring orders were produced on time.
(d) Carrying out such disciplinary procedures as were warranted using methods in the *Works Handbook*
(e) Keeping the factory clean and making sure that Health and Safety rules were rigorously enforced.

The nature of a first-line supervisor's job is its immediacy. He has to handle problems arising during his shift and make sure his next colleague has no residual problems to handle. Perhaps, more than at any other management level, the first-line supervisor acts in the shortest possible time frame.

Consequently, performance monitoring has to be immediate as well.

The key document in starting to achieve the changes we wanted was a daily performance sheet (Fig. 5.1). Simple though this now seems, it was a significant improvement on anything that had gone before, in ensuring that first-line supervisors thought about their tasks and their achievement.

The alternative of shop-floor data collection was expensive at the time and allowed no commentary on achievement to be made. Our system was simple and cost-effective in promoting keen, daily assessments of performance and what had been done to improve it. Production management is an unrelenting, unforgiving, challenging, achievement-orientated activity. Daily reporting as shown in the Figure enforces the necessary rigour for the task. The sheets were seen by most senior managers. First-line supervisors know that they are under scrutiny, their achievements to be praised and criticized.

The reports became the base for all management accounting reporting. Standard output rates and hours were compared with those achieved. Corrective action where necessary was listed. Each week the standard hours and actual hours spent were converted into monetary terms and an inquest held on who had lost money—and why. For the first time for many years, first-line supervision was put under sustained but constructive pressure.

124 Work organization and training [Ch. 5

SECTION MANAGERS' REPORT

FOR DAY _____ DATE _____

SECTION _____ WEEK NO. _____

MANAGERS _____

1. CAPACITY/PLANNED OUTPUT

MONITORING POINT	SOLUTION SPRAY	DANIELS	CAR PAD PRESSING	LANGLEY BAKE
5-Day Capacity Planned output Manning Normal hours Output/man hours				

Signed _____
Production Manager

2. CURRENT WEEK'S ACHIEVEMENT

MONITORING POINT	SOLUTION SPRAY	DANIELS	CAR PAD PRESSING	LANGLEY BAKE
Daily requirement				
Day's output				
Output to date (current week)				
Absenteeism (hours lost)				
Overtime (hours)				
Output/man hour (week)				

3. CORRECTIVE ACTION

4. REASONS FOR OVERTIME

5. REJECT REPORT

6. OUT-OF-STOCK ITEMS

7. GENERAL COMMENTS

Signed _____
Section Manager

Fig. 5.1. Daily performance sheet.

This activity, more that anything else, re-kindled the enthusiasm of first-line supervision, and made them think about their role and the need to include their workforce in their thinking. It made them target-conscious and better managers in just about every respect within their job. Without it, we did not believe that the cultural framework or foundation on which to build new working group organizations would be there and the whole basis for improvement in the factory would be missing.

5.6.3 First-line supervision in the 1990s
It is my firm contention that, without effective first-line supervisors, running a factory in the UK in the 1990s will be practically impossible. They are the bedrock on which technical, organizational, motivational and systems changes can take place.

Given the need for zero-defects, a vast overhaul of quality assurance procedures, technological change of a high order and changes in culture and work organization which would have been out of the question ten years ago, what kind of first line supervision is needed in the 1990s?

5.6.3.1 Redundancy
There is a strong possibility that some people who have survived the inertia of the seventies and the traumas of the eighties will no longer cope. They will have lost drive, ambition, perception of the changes they need to make in their own role. They will lack the innate ability to be re-trained and re-motivated. We found that, however hurtful, these people had to be changed. Redundancy was the final option.

5.6.3.2 Recruitment
In the past, our first-line supervisors were largely recruited from the shop floor. This no longer seems a fair requirement for most, perhaps even for a minority, of first-line managers. Whether it is possible to recruit graduates is questionable, but 'outside recruitment' of reasonably well educated people is essential. These outsiders should help to break any lingering cosy relationships between supervision and operatives. Cutting off a potential promotion escape route for operatives, however, could be counter-productive. If shop floor operatives are recruited, they should prove worthy of it, through study and attitude.

5.6.3.3 Pay
Pay has to be adjusted to take account of the required gap between reasonably high earners and the managers, looking after them. It is not necessary to close the gap completely , but it has to be narrower than occasionally exists now.

5.6.3.4 Discipline
First-line supervision needs to be supported on most, if not all, disciplinary occasions. Running a disciplined factory, when society generally is ill-disciplined, can be very difficult, if someone does not make rules and enforce them. I cannot remember a case where managers followed the recorded disciplinary procedures to the letter and did not eventually achieve the requisite outcome.

'Ill discipline' is often concerned with trying to 'circumvent the system', such as an attempt to be paid for things not done. Physical violence or a blank refusal to carry out a legitimate request are not major problems, but both do occur. However, in any self-recording bonus system, work-booking fiddles are usual if not inevitable. Initially we had occasions where our section managers signed as correct work-booking cards which, with even a cursory glance, could have been seen to be faulty.

Other disciplinary problems are 'rest times' and good housekeeping on the shop floor. Within standard PBR schemes the 'rest allowances' are a traditional way of ensuring that an operative can take quite a considerable time away from his job if he wants to. 'Slack rates' ensure this.

The answer lies not so much in chasing operatives to work harder and applying all possible disciplinary procedures, but in changing the payment system. Otherwise, machines should dominate the operative and not the other way round. If this is not possible and neither is immediate change in the payment system, rest allowances and rest rooms or centres need to be more carefully established than is normally the case. In Ford, the rest centres are located so that they can be seen by track supervision at all times.

A well-documented disciplinary procedure, probably following the ACAS code of practice, is needed. Once this has been recorded, all managers at every level should be trained in its use and instructed, no matter how laborious it seems, to follow the procedure totally. Only in this way will it be possible to ensure that management will win if a disciplinary case ends up in an industrial tribunal.

To re-inforce our own training activities, several well-publicized disciplinary cases followed, where it was proved conclusively that senior management backed first-line supervision and that the procedures worked.

5.6.3.5 First-line supervision cohesion and team working
Any rationally minded manager might doubt whether abseiling down a cliff face in Wirksworth Quarries or trying to find the site of a map reference in the mist-enshrouded, rain-washed hills of North Derbyshire could help in establishing social and role cohesion, extended networks and successful team working. But it did.

We took, as our starting point, the view that management performance can be improved significantly by building group identity and the self-confidence of individual managers. Action-centred leadership proved to be an ideal way of achieving these aims.

At the same time, the basics of team working were explored both in the training centre and on the rain-swept moors. This helped to:
- identify team objectives
- analyse team strengths and weaknesses
- ensure that a logical and agreed method of achieving objectives emerged
- improve working skills
- promote discussion on leadership styles and what appeared to be the most effective one.

All of these factors needed to be explored and understood, and some belief needed to be established that they could be replicated on the shop floor, before any successful changes to factory work organizations could be made.

5.6.3.6 *Training and team working*
While the action-centred leadership courses proved positive in gaining confidence and experience of team working in structured problem solving, training in all the other role activities of a team leader needed to be put in place. These are shown in Fig. 5.2 and described below.

(a) Size, composition, nature, culture of the teams
Teams or sections tend to grow up in all factories, either as a distinct part of the production process or, perhaps, even as a cost centre designated by the Management Accountant. An early review of all such centres, sections or teams is essential, to ensure that local supervision is capable of organizing the team and that it can be a satisfactory unit. Other factors which will need clarification include:

- The role of the team—what the team must do to perform well.
- Team composition—the number of skilled, semi-skilled and unskilled people needed.
- How the team should be paid.
- What service functions are needed and whether these should share some of a team's gains.
- What resources are needed to carry out the team's role—equipment, material, spares, energy, etc.
- Who does what and how.
- What hours the team should work.
- What flexibility within the team is to be permitted or encouraged.
- Training.
- Manning levels.
- Who solves problems, who relates to other teams.
- Who reports to senior management and how.
- What methods of working should be introduced.
- Who handles machine set-ups and breakdowns.
- What role should the local shop steward have—if any.
- What information collection and dissemination systems are needed.

Participation procedures—how these can be established and made effective.

(b) Monitoring performance of the team
The team's inputs and outputs need to be carefully defined. Suitable controls need to be in place. The 'real' performance of the teams should be:

$$\frac{\text{Value of output} - \text{Value of input}}{\text{Costs incurred}}$$

We made sure that the daily section managers' reports were reviewed and re-designed so that they:

(i) Planned and monitored all activities under the control of the section manager, not just the key items.

Sec. 5.6] First-line supervision 129

SECTION	SECTION MANAGER	CONFIRMATION OF SECTION	PERFORMANCE MONITORING & FINANCE TRAINING	TEAM BUILDING		INCENTIVES	SPC	PLANNING/ SCHEDULING	UNION ARRANGEMENTS/ DISCIPLINE	HEALTH & SAFETY
				ACL	GENERAL TEAM BUILDING					
DISC BRAKE PADS										
Cells 1–6	M. J. Smart	23 employees per shift plus 3 Tradesmen		Nov. '89			SPC Comp.	MRP II Comp.	Oct. '89	Comp.
	S. Campbell			Comp.			SPC Comp.	MRP II Comp.		
	B. Illingworth			Nov. '89			SPC Comp.		Oct. '89	
Cells 8–13	M. Cattling	23 employees per shift plus 2 Tradesmen		Comp.			SPC Comp.	MRP II Comp.	Oct. '89	Comp.
	B. Parton			Nov. '89			SPC Comp.	MRP II Comp.	Oct. '89	
	J. Canham			Nov. '89			SPC Comp.	MRP II Comp.	Oct. '89	
Cells 25–28 15–16	B. Cooper	23 employees per shift plus 2 Tradesmen		Nov. '89				MRP II Comp.	Oct. '89	
	J. Lynch			Nov. '89	Completed		SPC Comp.	MRP II Comp.	Oct. '89	
	S. Davies			Comp.				MRP II Comp.	Oct. '89	
	E. Whiteley			Nov. '89			SPC Comp.			
C. V. Manufacturing Area	J. Bailey	13 employees per shift plus 1 Tradesman		Comp.				MRP II Comp.	Oct. '89	Comp.
	P. Willett			Comp.				MRP II Comp.	Oct. '89	Comp.
	J. Collier			Comp.				MRP II Comp.	Oct. '89	
Jobbing/Competition Materials	J. Lindop	8 employees per shift		Nov. '80			SPC Comp.	MRP II Comp.	Oct. '89	
	M. Cronin			Nov. '89	Completed			MRP II Comp.	Oct. '89	
	S. Wheelock			Comp.			SPC Comp.		Oct. '89	

Fig. 5.2. Section manager training and general status.

(ii) Made it possible to provide weekly cost and variation statements for the teams. The statements in turn formed part of the overall performance statements needed by senior management.
(iii) Provided a full explanation for the out-of-course events which had taken place.
(iv) Showed team objectives as clearly as possible so that they could be understood by everyone in the team.

(c) Team building
Initially we re-considered the role of first-line supervisor. From a manager (we thought) he should become a facilitator helping his team to achieve team objectives, by ensuring that requisite resources were always available, by problem solving and by ensuring a smooth interface with other teams and functions. In retrospect, this was an over-idealistic way of considering the role of first-line supervision and we quickly abandoned the idea. It brought a dichotomy between the role of the manager as a team facilitator and the pressures put on him to achieve management objectives. Add to this team members who still have not been weaned away from believing that simple, individual PBR schemes were all that was needed to pay themselves. The gulf was unbridgeable at that time.

Any team needs a captain, and even if there is a large measure of concensus and flexibility, someone must be the team spokesman and planner. A first-line supervisor must still be responsible for the team achieving its objectives. He must still be a manager, no matter how much his team works as a team.

Team building can be alien to many managers schooled in a past bureaucratic and autocratic culture. Team building needs to be explored and explained.

(d) Incentives/payment systems
The question of payment has already been raised. What payment systems are relevant within a team activity needs to be carefully determined. Obviously group payment is important, but how this is calculated and then divided up among team members can be a problem. Whether it is done on a good-products-only produced, system-based, standard cost achieved against cost incurred, or some other method, can cause endless debate.

(e) Statistical process control
With increasing emphasis on zero-defects, the processes whereby these can be approached, if not achieved, need to be understood by everyone. At the same time, if operatives are to inherit the quality assurance function, then they too need to understand the techniques of quality control and how defects can, largely, be eliminated. SPC is a good starting point.

(f) Planning/scheduling
Establishing MRP II is a major activity and could have a significant influence throughout the company. Its potential effects need to be known by everyone in the organization. Any production team should want to know:-

- the order load—now and in the future
- the work content of the load
- order number and progress
- priorities
- achievements
- availability of raw materials and components.

MRP II should help in providing this information if reasonably efficient shop-scheduling is in operation, but anything which is as all-pervading as MRP II needs to be known and thoroughly understood by the team.

(g) Union agreement/discipline
Discipline has already been discussed.

The role of shop stewards and the place of union agreements need to be overt. Like the Personnel function, shop stewards may feel that they have a vested interest in keeping out-of-date practices in incentives and relationships going as long as possible. It is quite likely that many pay problems and other related grievances can be eliminated through team working if the elimination of relative deprivation is added. The traditional shop steward role should diminish. However, it is possible that a more relevant shop steward activity might arise in determining and resolving inter-team relationships and problems.

(h) Health and safety
The 1990s are likely to be a 'green' decade. Polluting the local river or the general atmosphere or indeed dumping waste without considering every consequence is going to be of increasing importance. Factories, too, need to be healthy and well-lit, the kind of places that the most fastidious grandmother would gladly walk around. There is no excuse for allowing operatives to work in potentially dangerous conditions of any kind. The first-line supervisor should know all the factors surrounding health and safety, especially the COSHH regulations.

(i) The Japanese way
Like many others, Ferodo's managers made the pilgrimage to the Nissan plant in Sunderland. They came back converts—to non-clocking; everyone using the same car park; common uniform for all personnel on site irrespective of position.

What impressed them most, however, was the standard of first-line supervision and how effective it was. British industry has some way to go to be just as good.

5.6.4 Conclusion
Without some restitution or perhaps even the new establishment of the status and skills of first-line supervision, cultural change and the effective use of teams on the shop floor will not come about. Without effective leadership, whether this is in management of the old style or in a modified facilitator role, teams will not work properly.

As a basic truism, the more responsibility you give to people the more responsible they become. This seems so with first-line supervision and team working.

5.7 THE ROAD TO SETTING UP NEW WORK ORGANIZATIONS

Several of the actions needed to be taken before effective team working or new work organizations are put in place have been recorded already. Others include:

(a) Job education
Any move towards an integration of company gradings and a common payment system will need a revision of all jobs and responsibilities. Essentially, this will mean a company-wide job evaluation activity in which all union groups will need to take part. Only through job evaluation will it be possible to have a common basis for relativities, which reflect the value and worth of all jobs being done.

Establishing 'bench marks' will be the initial concern.

(b) Payment systems
Initially the way to get Union agreement on changing pay systems might be to:

— Retain current pay levels and systems, but allow them to wither slowly, while the second tier takes over.
— Introduce a second-tier payment for the achievements of the working group as a whole.
— Operate a third-tier system based on product line, site or even company performance.
— 'Red circle' jobs where people are being paid far more than job evaluation believes is necessary will need to be bought out.
— A common payment system must eventually apply for everyone.
— Equal pay must be given for work of equal value.
— 'Earnings protection' will be a key union–management debating point in discussions on pay systems.

(c) Harmonization of pay and conditions
To have groups of people in which individuals have different conditions of employment is invidious. The cost of 'levelling up' could be considerable and will need to be earned from the improvements in productivity which new work organizations are likely to bring about. This again might cause some tense moments in discussion with the Unions. Precisely what harmonization of conditions means will have to be debated. Similar sick and pension schemes will probably need considerable funding. Abolishing of clocking-on could result in some quite severe time-keeping problems.
(Items a, b, and c are discussed further in Chapter 7)

(d) Union negotiation rights
In a company with a variety of Unions, separate negotiating procedures will be in place. With harmonization of pay and conditions, the need for separate negotiating rights will disappear. Getting the Unions to agree with this could be a problem.

(e) Redundancies
Any major change in work organization could result in redundancies. A statement that there will be no enforced redundancies might be useful in selling organizational change.

(f) Communication
In any project as far-reaching and as important to individuals as changes in work organization and associated practice, good communications are vital. So are weekly briefings recorded on 'pin boards', with personal briefings given to first-line supervisors who then brief their operatives.

(g) Management/Union agreements
It might be interesting to record the management/Union position before we introduced significant changes in work organization. A joint statement read like this:

(i) There will be no compulsory redundancies as a result of the proposed changes. Natural wastage and voluntary redundancy will handle any job surplus.
(ii) No employee will be expected to take a pay cut, without compensation.
(iii) Negotiating rights of the individual unions will be maintained but every effort will be made to institute joint arrangements.
(iv) Job evaluation does not commit anyone to a particular payment level or indeed a payment structure.
(v) No one will have to learn new skills, but anyone who does not cannot expect improved remuneration.
(vi) Job times may be changed, but major change will be in eliminating those things which prevent improvement in productivity from taking place.
(vii) Job grading could well change rates.
(viii) Single status/ harmonization will mean commonality in pensions, sick pay, periods of notice, holidays, pay relationships and methods of payment.
(ix) As far as possible there will be improvements in all services and activities which support working groups, especially those which the groups will not control. These will include the availability and quality of raw materials, components and maintenance. Factory planning will be improved.
(x) Some form of time/work study will continue as production data will still be needed for costing/accounting and measuring machine output.
(xi) Likely team members will be direct operatives, local maintenance and engineering personnel, factory schedulers, anyone directly concerned with moving material or quality assurance, plus section managers, who will still organize the teams.
(xii) Employee representation will be accepted at all levels of the change discussions.
(xiii) Special arrangements including transitional pay will be made where a person's earnings under a new scheme will be lower than those currently achieved.
(xiv) The emphasis, at all times, will be on working more effectively rather than harder.

5.8 TRADE UNIONS

Throughout this book, reference is made to trade unions and shop stewards and their potential for either a malign or benign influence on a company. It would perhaps be useful to draw on experience to comment more widely on trade unions.

Senior managers in manufacturing all know of occasions when companies have been coerced, to the point of collapse, by trade unionists, fully supported by their local national officers. Even in the last year or so, when trade unions should have been tamed and educated by government legislation, workers at the Birds Eye factory on Merseyside can apparently allow their factory to be closed down rather than accept cost reduction activities which would have cost some jobs.

From time to time, I received information from the Economic League (7 Wine Office Court, Fleet St, London EC4A 3BY). This organization sets out to show that industry must be constantly on guard against revolutionaries who use trade unions as a wooden horse, from which they will jump out and destroy profit-making units. Politically motivated disruption threatens every company in every branch of industry.

Michael Edwards, in his British Leyland days, might agree.

There is a more general view that unions have used physical intimidation to obtain unfair trading advantages. They have used their monopoly power to exploit their employers. The 'enemy within' is to be mistrusted.

Yet, in Germany, the union I.G. Metall is often accused of the same tactics. In Italy too, a strong Union movement has won major advantages for the workforce.

So, do trade unions 'impose restrictive practices, force up pay, deter investment, hold back productivity growth, ultimately destroying jobs? A report in the *Financial Times* of 29 August 1989 suggests otherwise. The conclusions, it says are at odds with many popular views of unions. For example, productivity growth was as good in unionized companies as in non-unionized. Not much of the increased productivity at companies with trade unions was due to government legislation.

The following is probably an oversimplified summary of several reports, but bears out my own experience:

(1) Government legislation had little influence on trade union co-operation, power or ability to restrict productive improvements.
(2) The shock of near-bankruptcy was an effective catalyst in creating better relationships. There were some production managers who believed that 'We should have taken the unions on more, when we had the chance'. Revenge is a pretty poor reason for taking away well-established practices and procedures.
(3) Irrespective of government legislation, negotiation of flexible working agreements, team organisation, multi-skilling or basic productivity needs careful, patient, sustained negotiation with some inducement thrown in.
(4) Our unions did not want to 'take-on-the-management' or become militant in their demands for pay, once the company trading position had been explained carefully—time and time again.
(5) National officers, who rarely took part in local briefings and depended upon shop stewards to inform them of company finances, often took a different stand. The union head office had obviously established some sort of a norm for the

pay round. The statement that 'X company ten miles away had settled for $7\frac{1}{2}\%$ and so therefore should you' was often used as a bargaining ploy. We ignored this kind of remark.

We told our local union what the company could afford and paid accordingly. However we left the option open, for them to help us with ways and means of improving productivity, and the benefits in which they would share.

National officers can be destructive of local cooperation no matter how hard won it may be.

(6) The great curse of many British plants is to have three or even four union groups, battling to represent some part of the total workforce, often attempting to leap-frog over each other in terms of pay and conditions. It makes life very difficult when one group has severe relative deprivation. Harmonization of conditions will help, but even getting this started could be difficult.

(7) Individual shop stewards are often a problem. The puzzle for management is to determine how far the 'bolshie' shop steward truly represents his members. Will they follow him or a management which is seen to be fair and honest? The solution here is to make sure that all shop stewards and especially the most militant are exposed to as much management thinking as possible and as many presentations of monthly, quarterly and annual results as possible. Understanding brings some if not complete cooperation.

How far should management appeal over the heads of shop stewards? The answer is, totally, if necessary. A major cause for concern must be that shop stewards are allowed or even told to tell their members what is going on, what plans are being made, what results are being achieved, what prospects for future employment there are. This is a management activity and must be done by management and given a management viewpoint.

(8) Employees have a greater potential to damage their company now than at any time before. Once, stocks at Ferodo were so high that an overtime ban for two or three weeks over some dispute had a scarcely noticeable effect. With much lower stocks and contracts in Germany at risk, a stoppage would be a disaster. So stoppages must be avoided—by constantly telling people what is going on, by ensuring everyone knows that a stoppage will cost jobs—their jobs—and making part of the pay earned consequent on achieving good performance consistently.

(9) The failure of British trade unions lies in not defending jobs sufficiently well, not in failing to get more pay. Once an employee has been made redundant, the union forgets him on her. Those who remain in employment are the ones who will pay union dues and so must be supported.

A major part of education within the unions should have been how union officials and shop stewards can make companies more efficient, more competitive and more likely to increase employment. Prosperity should begin on the shop floor with debates about raising productivity. They should have been pushing management to improve company performance.

Management should help to correct this situation by pushing and aiding shop-steward involvement in discussing ways and means of improving company profitability.

(10) In Germany, with a much more structured participation process the unions have been more cooperative, but even so militancy has not been eliminated. The same process of working together to achieve corporate goals is needed.
(11) Trade unions can be both negative and positive factors. What they are, largely depends on management, not the Communist party or any other quasi-revolutionary organization.

5.9 DISCIPLINE/UNFAIR DISMISSAL

Discipline and how it can affect first-line supervision was considered in section 5.6.3.4. There is a broader view which it is necessary to record.

Any senior manager is likely at some time to get involved with an unfair dismissal case. Regretfully, we found that both the people concerned and shop stewards would demand to go right to the end of disciplinary procedures, no matter how justified the disciplinary judgement appeared to be. The obvious conclusion was that the accused person had nothing to lose by doing so and that the Trade Union could not be seen not to be defending one of its members.

(1) We learned a lot from *Rigby v. Ferodo*. Never again would we operate without every last agreement and procedure being recorded.
(2) The code of practice we had in place needed to be understood by everyone and followed to the letter by the managers concerned. We had plenty of practice in this.
(3) Managers tended to see ill-discipline and take action far more readily than the company's Personnel function thought appropriate. The two groups have often conflicting objectives. The Personnel function sees itself more as an arbitration activity, promoting harmony and ending conflict. Line managers, pushed to achieve production targets at a specific cost, demand that operatives are well geared to help in this task. The odd punch-up does not often concern them. Hence the dichotomy of views in disciplinary procedure between line managers and the Personnel function.
(4) The stage before the action goes to an industrial tribunal is labelled 'The final appeal'. It is then that the regional organizer of the trade union and the most senior manager available (usually a director) re-hear the disciplinary activity. If local management has followed the code of practice to the letter there should be no problem.
(5) For the company to avoid a verdict of unfair dismissal it is necessary to prove:

 – The dismissed employee was given a fair hearing.
 – The evidence produced by the company against the employee did warrant dismissal, and not perhaps a lesser punishment.
 – The proved misconduct was great enough to warrant dismissal.

(6) In our case we tended to build up 'case law', so everyone knew what lapses in discipline produced what result. For example, work-booking offences, when found, always resulted in dismissal.

(7) Often, management may be well served by demanding a 'pre-hearing' if a dismissed employee opts to go to a tribunal. This is a prior, legal view, carried out to assess the chances of success or failure of the case at a full tribunal. If management 'wins' the pre-hearing, it is usually stipulated that if the union side still supports the case at tribunal, they could be responsible for quite major costs. This normally ends the proceedings.

(8) The final appeal stage is an occasion when all possible evidence is pulled together and presented in as clear and precise a way as possible. The summing up should be recorded carefully and set out as a prime document for a tribunal. The evidence from both sides needs to be recorded together with the reason why one side is accepted and the other rejected.

(9) There is no reason why any reasonably competent management should not have a clear-cut disciplinary code which they act upon when necessary. Most people will accept it gladly. Management can do considerable self-injury by not keeping to rigorous disciplinary procedures.

The only occasion we had, when it was obvious a major disciplinary procedure should have been activated and was not, resulted subsequently in a further failure in discipline which led to a tribunal hearing.

(10) *Note*: Unfair dismissal is covered in the UK by Section 57 of the Employment Protection Consolidation Act 1978, as amended by section 6 of the Employment Act 1980.

5.10 TRAINING

5.10.1 General

Few companies seem to train well. Few train sufficiently. Few have a training strategy. When I became director of a site with a turnover of £55 million, no one ever said that I should be trained in a particular subject.

For many years we trained in the traditional way. A manager or supervisor was sent to an outside agency for two or three days, though occasionally for a longer period. Once the course was over, we expected that 'trained' person to come back energized with the new knowledge and immediately make an impact on his job and the company generally.

It seldom worked out like that. Unless the person is of reasonable status, extremely strong-minded or determined, or has a work organization which will accept influence readily, then the training must be of minimal value.

Of course, for apprentices and other fairly junior people we funded day release and other technical courses. Managers, however, often believe that they cannot spare the time to be trained or perhaps retrained. They tend to dislike working at home. Distance learning via the Open University and elsewhere is excellent, if managers in mid-career will only take it seriously.

The least effective way to train is to answer a cold mailshot. A need should have been worked out for training and the best people to do it a long time before one of these is taken up.

5.10.2. Why train?

We decided that there were five reasons why training was necessary.

(a) To create organizational cohesiveness. For example, to ensure that the 'train principle' worked and that managers at all levels knew what the company methodology was, what key objectives were extant, what courses of action needed to be followed, and what individual managers had to do without being told to do it. An early indication that many of our managers were 'financially' illiterate brought forth financial, costing and profit-planning training.
(b) To improve or indeed establish effective work organizations which would help to energize company performance. The training we gave to first-line supervision was part of the process.
(c) To ensure that there was a company-wide knowledge of major developments e.g. MRP II, SPC, Health and Safety, etc.
(d) To enhance or perhaps gain specific and needed skills in the company for specialist activities—electronics, some aspects of computing etc.
(e) To generate or perhaps re-generate people with requisite technical skills, starting with apprentices and probably ending with fifty-year-old members of the engineering work force who needed new engineering skills, especially multi-skilling.

For the first three applications, outside training is not suitable. It does not provide courses based on relevant internal experience, nor does it train enough people at one time.

For the fifth type, it is likely that the local technical college can be persuaded to establish courses which can directly relate to company requirements.

Internally generated courses provide the fastest, most tailor-made training possible. Any number of participants can be handled.

The more that take part, the cheaper the training becomes.

Two of the most important courses we undertook were in Finance and Profit Planning.

The glue binding all activities in a manufacturing company is the need for all people, in all activities and functions, to perform well. The measuring of performance needs and results is best done in financial terms.

We had been particularly bad at giving out financial information in a way which made sense. Once, when I asked some question on the accounts of the then Financial Director, he retrieved them from the safe in his office and made sure that I could not see anything other than the data I was querying.

In the early eighties, management was reducing costs, making people redundant and trying to generate cash, all without any fundamental understanding of the financial results of their actions.

The situation for the shop stewards was even worse. How could they trust data they did not understand? How could they agree to proposals when they did not know whether they were realistic, valid, needed or not? They had no knowledge on which to base realistic alternatives to what management proposed.

Over and over again, UK trade unionists appear to look at the final profit a company makes and base their wage demands accordingly. No one tells them that return on investment is still only 10% or that cash is running out; or if anyone does, they do not believe what they are told.

How often has a UK management said in a pay dispute: 'This is all we can offer. If we give you any more there will be serious repercussions including redundancies'. Then a further offer *is* made and *no* redundancies take place.

There are some exceptions. Ford in 1990 raised final pay offer, but then reduced capital investment in the Bridgend engine plant.

The answer, we thought, to unsustainable wage demands and pay rates was to teach as many people as possible about finance.

The process has to be long and wearying. The need to make profit, and its relationship with new investment and therefore job creation and security, has to be spelled out carefully. The more profit plans are debated and the more monthly results are explained, the greater the potential understanding and eventual co-operation of the workforce.

We began in 1982 (too late by at least 10 years) on a process of financial education which would cover directors, managers, supervisors, specialists and shop stewards alike. We also decided that the course should relate to the company's past, current and future performances. We made the company a case study, showing in financial terms what went wrong, what was done to put it right, and what still needed to be done to ensure future prosperity.

Knowledge of the company can be linked with financial theory. Beliefs, feelings, generalization could all be tested by hard financial reality.

The course[†] produced a cohesiveness in management and a trust between management and shop stewards which no other procedure or activity could have done. *If any single course of action converted an organization which might have died on its feet to a manufacturing phoenix, it was this one.* For the first time, managers and shop stewards spoke the same financial language. Trust developed. No one in the future could ever say, 'Nobody told me' or 'If only I had known'.

A summary of the course as originally set up is given in Appendix 3. The data used to analyse company performance is also given.

5.10.3 Communication

Probably as a concomitant of the finance/profit-planning courses, Ferodo was brave enough to tell everyone what the financial results being achieved were, far earlier than most other manufacturing companies. Good, bad and indifferent results alike were told, along with anything else which seemed important. No one at Chapel ever misused the information, either in putting in excessive wage claims or in leaking information to competitors. The bravery paid off.

The combination of finance courses and improved communication was part of the same process. One without the other would not have been effective.

[†] The course was designed with the help of the Loughborough University and David Barnes and Associates, Finance House, Trelawny Square, Flint, Clwyd.

Five years after starting finance training the first union group decided, rather hesitatingly, to accept performance-related pay. The timescale seems inordinately long, but where a company has only modest trust between union and management, the time taken to establish confidence can be measured in years rather than in months. There is no such thing as a 'quick fix' in UK manufacturing companies.

The Chapel-site communications programme worked like this:

Timescale	Personnel involved	Activity
(1) October	Senior shop steward of each Union group and senior management	A meeting to show and then discuss the putative profit plan, explaining amongst other data: • Potential revenue • Capacity utilization/forecast and manning levels • Pay rises • Other forecast costs • Potential profit • Key strategies • Potential investment
(2) December – January	Managers, supervisors, shop stewards	Two-day course to debate the completed plan, with no relevant data refused. Copies of the plan issued, but returned after use. Action plans to achieve overall plan agreed. Achievability of plan tested and necessary changes agreed. Economic forecast for the year discussed
(3) Each month, following the Board Meeting	Managers—followed by a briefing-down by managers to their subordinates	Monthly results. Financial/Management Accounting information, showing revenue earned, cost contribution, operating profit. Analysis of what went right, what went wrong and what is likely in the future
(4) Quarterly	Managers	Continuation of (3), but with a review of the profit plan action plans

(5) Monthly	Shop stewards and local site director and managers		Monthly reporting details released; approximate to the meeting with managers but with greater emphasis on matters directly concerning unions, output rates, manning, overtime, etc.
(6) Quarterly	All people on the site		Quarterly results. A quarterly resumé of revenue, gross margin, profit and forecast. Written review made by site director and adapted where necessary by local managers. Production stopped and operatives paid to listen to the report.
(7) Weekly/ monthly	All operatives		Weekly/monthly data sheets, prepared by local production manager and issued to all operatives—mainly concerned with output rates, rejects, investment, current profit earned by product lines, etc.
(8) Annually	All personnel		Annual T&N/Ferodo Reports

5.11 SENIOR AND MIDDLE MANAGEMENT TRAINING

Like many other UK manufacturing companies, we had a core of mature middle managers with limited qualifications, yet carrying out vital company functions. These people test any training activity to the full. (By far the most important element is not, perhaps, training as such, but the role model set by directors, and ensuring that this is followed.) We undertook the following internal courses:

(1) Finance and Profit Planning, covering:
  Financial control
  Budgeting
  Top-down planning
  Profit planning
  Strategy formulation
  Marketing
  Objective setting
  Project management

(2) MRP II, covering:
    MRP/capacity planning
    Shop scheduling
    Purchasing
    Information technology

(3) Business Improvement Planning, covering:
    Work organization
    Motivation
    Group training and incentives
    Job evaluation
    Presentational skills
    Assertiveness

(4) Quality Assurance, covering:
    Statistical process control.

All training was directed towards achieving comparatively similar levels of knowledge across a broad spectrum of managers, creating a common technical base for interfunction and job relationships. Organizational cohesion was the real goal.

5.12 TRAINING FOR MULTI−SKILLING

Once it took five years to train a craftsman—fitters, turners, electricians, joiners. Many of these crafts grow increasingly obsolete, if not totally redundant. With current demographic trends showing that there will be fewer and fewer youngsters available to take up apprenticeships even if they wanted to do so, something of a crisis could occur in skill availability. The time has come to consider multi-skilling.

The advocates of multi-skilling generally concentrate on those jobs or activities which were basically done by 'craftsmen' They suggest that the introduction of new technology and work organizations associated with it demand new skills and attitudes. 'Multi-skilling' provides the answer.

The emphasis on 'engineering staff only' starts to uncover part of the problem. Multi-skills need to be in place right across the company. Gone should be the days when a salesman was a salesman only, or accounting was left to the accountants, or personnel work to the Personnel function.

What are the motivating factors which are pushing the acquisition of multi-skills everywhere? These are the main ones:

(a) Any manufacturing company faced with serious competition has to raise its productivity on a year-by-year basis. This will inevitably entail raising output while using the same people, or keeping the same output and reducing people. Inevitably, the ratio of people and output will need to change.
(b) If it is expected that fewer people are needed, the breadth of jobs must widen. The need to be more flexible will dictate that a much wider range of job skills will be needed to cope.

(c) Despite unemployment, most companies suffer from skill shortages of some kind—computer specialists, electronic engineers, material scientists. Recruiting such people is rarely easy. The pay they demand is often higher than that currently paid to similarly qualified people in the company. By far the best solution is to re-train internal staff who have shown aptitude and desire to become multi-skilled.
(d) The need to become increasingly competitive entails the much wider use of techniques, philosophies and technologically superior equipment. Inevitably, if local people cannot adapt or change and gain the basis for improving the company's competitive position, life will grow harder for everyone.
(e) UK line managers have relied heavily on specialist services to provide them with technical help—management accountants, personnel managers, management services, computer specialists. Often this has produced a dependency of spirit and of decision-making which has undermined the position of the line managers.

Every manager should be his own 'human resources specialist', for that is what the job entails. Delegating pay and discipline, even health and safety, to someone else hinders more than it helps.
(f) British companies, at least, should recognize that they can no longer afford to have a whole panoply of staff personnel to help line managers make decisions they should be capable of making themselves. In Germany, in particular, staff functions are already comparatively unimportant, when compared to those in the UK.
(g) New work organization will only add to the need for multi-skilling. Once pay is largely determined by the difference between what material is input and what good products are output, it is surprising how quickly teams want to carry out their own specialist activities. Where teams have been told they can bid or not for extra maintenance or personnel services and have the cost deducted from their potential earnings, they quickly decide that they can do without. This situation can be dangerous if requisite skills are not integral with the team activity. It is too easy to say a machine can be maintained properly and then to see it ruined as a result of a lack of technical skill to do the maintenance.
(h) New work organizations are also enabling 'layers of management' to be taken out of the system. Establishing autonomous work groups will reflect on the need for managers—in numbers and skills; numbers will decline; skills need to be enhanced. If discipline and 'management' are not so important when teams become self-disciplined and manage themselves, what happens to the manager?
(i) At the 'maintenance-engineering level' there is possibly even more urgency to make multi-skills a major part of the operation, otherwise the obsolescence will grow apace. With the advant of CAD/CAM and wire-erosion, process-controlled tool-making, highly specialized tool-making could be a thing of the past. What happens?

Possibilities
(a) Multi-skilling has to be directed towards people in their middle years. This is contrary to all training directions in the past, where more-junior members of the organization have used up most of the limited training budget.

(b) It is unlikely that multi-skilling of the nature required can be done internally. How this is to be brought about needs careful debate. Many companies have the same problem.

The Training and Enterprise Councils and the Engineering Employers Federation should be used to encourage vocational training especially for those who are past 40 years of age.

Many local technical/polytechnic colleges have had some experience in multi-skilling activities. Again, they need encouraging to develop appropriate packages.

(c) The training budget should be raised considerably. Spending up to 3% of revenue on training now seems about right.

(d) Everyone in an organization needs to be trained or to be qualified in a core skill—electronics engineer, CAD/CAM technician, MRP II specialist. Then a whole variety of non-core skills should be added. In the case of managers, for example, these might be as shown in the table below.

Core skills	Non-core skills
(1) Materials Management and MRP II	• Management Accounting/Financial/Profit Planning systems • Computer systems design and application • TQM • Human Resources • Business Management
(2) Electronics Engineer	• Management Accounting/Profit Planning • Health and Safety • CIM/Computer software engineering • CAD/CAM • TQM
(3) Production Manager	• Management Accounting/Profit Planning • MRP II • CIM/Computer engineering and control • Human Resources • Business Management • Legislation
(4) Distribution Manager	• Management Accounting/Profit Planning • MRP II • Human Resources • Team Working • TQM

The non-core training should provide the basis for:

- coherence training across the company for corporate objectives and performance improvements;
- the diminution if not abolition of staff functions.

The industry–education relationship

Some of the Americans have concluded that it is not the actual lack of money within the education system that causes concern: it is largely how the money is spent and what end-product eventually emerges. American children spend 180 days per year at school, while the Japanese attend for 240 days. This factor is compounded by inequality between the already affluent and well-trained and those who are not.

Seen from a relatively senior position in industry, the UK educational system seems to demand more and more resources (especially to pay teachers more) and yet produces fewer and fewer people who have sympathy with industry and actually want to work in a manufacturing company.

If in the local school discipline is absent, pupils abuse teachers and play truant, how can a factory be run effectively when the schools's products come to work there?

A factory is a dangerous place for people with a drink or a drugs problem. Ill-discipline and general horse-play can create chaos. Self-certification of illness did little to improve discipline and absenteeism. The Personnel Department in its social welfare role might be positively harmful to discipline.

A criminal record should always be treated with suspicion, but it should not be a bar to employment. We fired people not because they had been to jail, but for not telling us of their past criminal record. Theft to some degree is endemic on a small scale in factories. It should not be force-fed by slack discpipline.

We still trained too little in the eighties and did too little to help to create a better industry–education climate through the educational system.

The uneducated and untrained are really competing for jobs with better people in South Korea and Taiwan. Wage levels have to be reduced accordingly—or if not, the company is no longer competitive.

This is a tragedy, not just for the companies concerned, but for society as a whole. It means, if not corrected, a continuation of the wholesale transfer of jobs out of the USA and UK to the Far East.

What can an individual company do to combat this situation?

(a) Links with local schools are essential. Schools should provide the needs of industry in terms of skills and general aptitudes. It is no longer good enough to let education stay in the hands of educationalists and local education authorities.

(b) Training and re-training in the basic skills, probably starting with literacy and numeracy is needed. Manufacturing industry has traditionally recruited those with minimal qualifications and abilities. As more high-tech equipment is introduced, these people are just not good enough.

(c) Schools and their pupils should be informed of the skills now needed and an indication of requirements established via the aptitude testing which companies

have been forced to adopt, even for filling their most low-grade occupations.
(d) It is not just technical training that is important. A whole range of attitudinal, group dynamics type of training is needed. If team working is going to be a key aspect of new work organization, the necessary skills in terms of flexibility, self and team motivation, social integration all need to be taught.
(e) If a company sets plans for the future at all, it should determine the kind of skill group it needs both to manage and to operate its plants in the future. Machine minders, labourers, simple material transfer people, even product assemblers, may be in serious decline.
(f) Harmonization of conditions across a manufacturing company should destroy the division between 'white' and 'blue' collar employees. One certain way to obtain organizational coherence, so as to make sure that everyone has similar opportunities, attitudes, and appreciation of what must be done, is to gain similar educational/training standards if this is possible. This must mean that everyone should take part in training and education, whether the person is from the shop floor or the boardroom.
(g) There is an obvious shortfall in highly skilled people, whether in science or engineering. Universities and colleges are not turning out enough of either. Companies might consider more sponsorship of suitable people, so long as there is a relevant reward at the end of the process.
(h) As the demographic situation worsens for manufacturing, more women and older people will have to be recruited. This may be no bad thing if loyalty is more apparent. Jobs will have to be engineered accordingly, with little or no heavy lifting. The days of heavy manual labour may be over. Working hours may have to become more flexible to cope. Older people, for example, may need to work only 3 or 4 days a week.
(i) With higher numbers of single-parent families, and shattered homes, deprived social backgrounds and poverty a perennial problem, many younger people coming into industry will be emotionally disturbed, perhaps irritable, even drug takers. How does the employer cope unless company social cohesion is fostered and an extended family of relationships built up?
(j) Teacher placements in industry should increase. They should gain respect and possibly higher pay, by spending a year doing a reasonably senior job in industry. If morale in the teaching profession is low, it is due in some respect to the latent hostility that many people in commerce and manufacturing have towards the teachers. They feel, perhaps rightly, that teachers and the education system have let them down. Teacher placements could help raise teacher morale, show them that education is very important and create a better understanding of problems in the world of work and education. Conversely, managers from industry might become part-time teachers.

CONTENTION

(1) An effective work organization is one where power, authority and responsibility have been reconciled.

(2) Hierarchies and line and staff are anachronistic in terms of establishing effective work organizations for the nineties.
(3) Some of the eighties prescriptions for putting organizational defects right, such as focus sharpening and quality circles, need to be regarded with some suspicion. They often fail.
(4) Solving goal conflict problems should largely be a matter of work organization. Goal conflict, even between senior managers and the company, is often inevitable to some degree.
(5) Effective organizations need to be based on flexibility, industrial pluralism, reconciliation of power, authority and responsibility, and minimization of goal conflict.
(6) Senior management should give a lead in abandoning obsolete privilege associated in the past with status and position. The company car should be abandoned except for those who need it for their jobs. If management continue to use *nil satis nisi optimum* (nothing but the best for us) as their maxim, conflict can only continue.
(7) Effective first-line supervision is essential if major improvements are to be achieved on the shop floor. First-line management should be trained in performance monitoring, team building, incentives, SPC, MRP II, union and industrial relations, and Health and Safety concerns, before any change in work organization is contemplated.
(8) Training—nobody does enough or does it well enough.

A suitable set of strategies

(1) We will continue to introduce organizational changes which make effective team units with appropriate autonomy for local resource utilization. Pay will be related to good output from the team for all team members.
(2) We will continue to change the culture of the company to ensure that everyone operates in a flexible, non-hierarchical fashion, free from claims on status and privilege.
(3) We will continue to make management more effective, especially first-line supervision, through training and the application of improved planning and control systems, operating on a shift/day basis.
(4) We will attempt, further, to eliminate relative deprivation. It is not pay rates as such which cause most industrial relations problems, but one individual's pay compared with another's.
(5) We will extend all training activities, especially those which are carried out internally and which are largely aimed at achieving organizational cohesiveness.
(6) Labour productivity will be improved further by:

   (a) introducing more and more team incentives with money paid for producing good, sellable products;
   (b) having an integrated grade and pay structure covering all employees below senior management;

(c) introducing training and culture changes which will encourage employees to learn new skills and to be flexible;
(d) introducing unified terms and conditions for all employees, regardless of union;
(e) minimizing use of Taylor-type 60/80 work-study-based incentives, which are tedious to introduce and, because of the rating element, rarely give an accurate measurement of performance;
(f) introducing production equipment which is much more dominant over production personnel than has been the case so far;
(g) ending pay anomalies which stifle increased performance.

6

Systems and information technology

6.1 INTRODUCTION

Information technology appears to defy an appropriate and useful definition. When some consultant, keen to enhance his fees, says the nation is falling behind in information technology, most people in manufacturing industry appear to remain unmoved. They, no doubt, believe that they already have enough information—somewhere. The problem arises in relating one fact with another and being able to manipulate the whole to make decision-making more effective.

There is, though, the keen managing director with his own powerful personal computer sitting on the edge of his desk, capable of showing him trends in profit, revenue, contribution, gross margin, variable cost, all in bar charts, graphs, pie diagrams. Then, by using a simple exponential smoothing forecasting technique, he might ponder on what might be the result at the end of the year.

This certainly is information, though in the form shown on the spread sheet it is scarcely more useful than if the management accountant had scribbled out similar figures on a piece of paper. Information, it used to be said, is power; but power for what purpose?

For a manager of a major company function there is nothing more frustrating than to receive a set of monthly accounting documents which shows a completely baffling profit performance. Output has been good, overtime minimal, rejects low, expenditure on maintenance and repairs well under budget, fixed costs no greater than they have been in preceding months, yet profit is practically non-existent. A mad scramble through all the figures can then ensue to, hopefully, determine what has gone wrong, what has been missed, what seems like an aberration, what corrective action is possible.

Strong abuse of the management accountant at this time is not usually beneficial. It is too late. Systems and information technology just have to be re-developed to give a better result in the future.

6.2 IT—DEFINITION AND USE

Information technology is the combined use of computers and communication technology to store, manipulate, produce and transmit information, both in and between organizations.

Like many other new developments (like computers initially!), IT has promised much but tended to let users and potential users down. It is too easy to say that 'IT could be the technological spine of the economic, social and cultural body of Western Europe',[†] but the relevance of the comment will escape most managers in UK industry.

The IT technologists will be searching for methods to link the various technical facets together. General managers will want to know how IT can actually help them run their businesses better. The chance that the specialists will become so obsessed with their technology that IT will fall into disrepute is quite high. Line managers should be wary.

One of Ferodo's first essays into computer/telecommunications linking was in putting the various depots throughout the UK on-line with the stock records held centrally, and with the other depots. Costs escalated very quickly and were beyond those anticipated. Very small parcels of stock began to be moved from depot to depot, as depot managers found how easy it was to access records and then move goods physically from one location to another, instead of waiting a day or two for another delivery from the factory.

However, some of the more important IT factors which need to be understood and where necessary applied are these:

(a) Communication and networking
Access to and transfer of information is likely to grow in importance. For many years, suppliers of word processors have tried the additional concept of 'networking', or local area networks. 'LAN' is software which links fairly disparate electronic equipment together, to permit the shared use of available information.

For a company using IBM or IBM compatible computers, systems application architecture (SAA) will become increasingly needed. This is IBM's attempt to enable users to run the three main IBM ranges of computers by accessing the same software.

(b) Information retrieval
Computer specialists have wrestled at length with file design and information retrieval systems, especially when data bases grow larger and stored information covers more and more functions and activities in the company. Irrespective of how it is stored, these functions need information given to them in a format they can use effectively.

[†] *I.T. Europa Management.* May/June 1990.

Information retrieval software development is concerned with the means of storing and retrieval of information so that functions receive exactly what they want, when they want it.

(c) EDI
EDI or electronic data interchange, is one of the applications which should make line managers believe that inter-organizational IT activities could be of major benefit. EDI should allow information to be transferred between companies, particularly things like orders and invoices. This, it is said, will reduce costs and errors, improve cash flow and competitiveness, get a better grip on stock and so enable it to be reduced, and improve customer relations at the same time.

EDI should make a very positive step towards the reduction in paperwork of all kinds. The key to its introduction lies in establishing common computer/telecommunication conventions between different organizations, perhaps different companies.

(d) Equipment—general
Whether it is in facsimile transmission, personal computers, teletext or word processors, IT technology and equipment is developing rapidly. Competition is increasing. Potential users should take the most up-to-date advice/information possible. This may mean using consultants.

(e) Teleworking
It has long been assumed that teleworking or working at some distance from the place where an activity is going on will eventually take over from working on-site. The data manipulation and communication technology is now largely available to allow teleworking. The benefits seem sound—lower absenteeism, increased productivity, lower office costs at 'Head Office'.

Two types of teleworker are developing:

(1) those who are employed to carry out relatively mundane tasks like data entry.
(2) specialists or technicians who need not be 'on site' to carry out part or perhaps the whole of their function.

Neighbourhood or local work centres could be set up. The equipment needed will include:

— the teleworker's equipment: perhaps a personal computer; simple, if a data entry only is being carried out;
— the central unit: a dial-in service seems a necessity;
— telecommunication system: usually a high-speed mode will be required, plus a fax facility.

As in all IT activities the system must be secure—if possible.

(f) Information
Information should strongly influence how the company is organized, e.g., how the variety of potentially contending managers and specialists can unite to form effective

working groups, and how the company relates to its customers. Organizations should become 'information-based.'

The danger lies in concentrating on technology and not its practicality. It always seems possible to send more and more data down a line, or improve the speed of writing software, but not to make sure that IT works for the benefit of users. A hard, demanding look at costs and benefit needs to be made, just as if it was a production machine being bought for the factory.

6.3 SYSTEMS DESIGN

The basis of all information technology should not be the equipment used or even the software, but the systems being processed. Without them, everything else is of little consequence. We thought the two systems which had (and needed to have) an all-pervading influence on company activities internally were:

- MRP II (Manufacturing Resource Planning)
- Management Accounting.

MRP II is a closed-loop system, obviously computer-based, which unites most if not all of the functions concerned with order processing and production planning, in order to produce the most effective method of utilizing production and other associated resources.

Some specialists believe that Management Accounting should form part of MRP II in the production of a Master Production Schedule. We never saw it in that light. In its top-down planning and control process, Management Accounting is a separate and important function in its own right.

We thought that systems design should follow these precepts:

(a) We should concentrate on those activities which were crucial to running the company. Any idea that an expensive Management Services Department should spend a predominant part of its time, say, changing the nominal ledger package or making some esoteric improvement to the export order processing system, should be quickly knocked on the head. We eventually chose to develop two systems: MRP II and Management Accounting. Everything else was considered to be of comparatively minor importance.
(b) The key to getting reasonable results is to establish a database which most systems can use—production capacities, product bills of material, standard cost of operations, etc. Database management might be a tricky, specialist data processing activity, but getting information for the database should be comparatively simple, if lengthy.
(c) Systems should be interactive, within their own modules and with users. They should provide a link with all managers in the company. The managing director should be able to access the work-in-progress file, to determine the likelihood of an order coming off the factory on time, just as a clerk answering a customer's enquiry should be able to do.

(d) Systems interaction should be built into systems design. For example, it should be possible to judge, say, the effect of accepting an order on:

- the order load file and priorities
- capacity utilization
- shop scheduling
- deliveries generally and this order in particular
- potential extra revenue, contribution, profit to be earned and when.

(e) Managers who use systems should be able to ask the 'What if we did this?' kind of question and determine trade-offs—the gains and losses involved in taking one action as against another.
(f) Systems should possibly provide an interface with other companies, both customers and suppliers. Linkages between one and the other should help to ensure that deliveries both inwards and outwards improve and stock reduces.
(g) Systems in a manufacturing company should help to drive the unit, if not on a minute-by-minute basis, then at least daily.
(h) What is required is often complex and beyond the capability of local data processing/systems personnel to design and program. Hence the need to use software packages or, in the case of MRP II, one big package. The difficulties of doing this often appear as great as using an in-house design, but are not so. How and what packages are chosen deserves considerable management time. These rules seem appropriate:

- Make sure that the software needs have been well stated.
- Test the market—see what is available.
- Discuss uses and usages with suppliers.
- Choose a solution which others have found comparatively easy to apply.
- Be prepared to use standard features and re-write as little as possible.
- Make sure that the software supplier will be able to support the package.
- The whole activity should be governed by a 'user specification'.

6.4 OPERATIONAL PLANNING

Before attempting to introduce MRP II we decided on an Operational Planning approach to developing the internal production planning system. The elements we considered constituted Operational Planning are shown in Fig. 6.1.

6.4.1 Conflicts and organization

The conflicting objectives of production and sales/marketing often clash in the planning/stock control activity. How the company is organized will often determine who wins the battle, of a limitless product range with infinite production flexibility and over-large stock, against minimum machine set-ups, maximum batch sizes and over-large work-in-progress.

By the late seventies, this kind of conflict had reached such a point in Ferodo that only a substantial organizational change could ensure a major improvement.

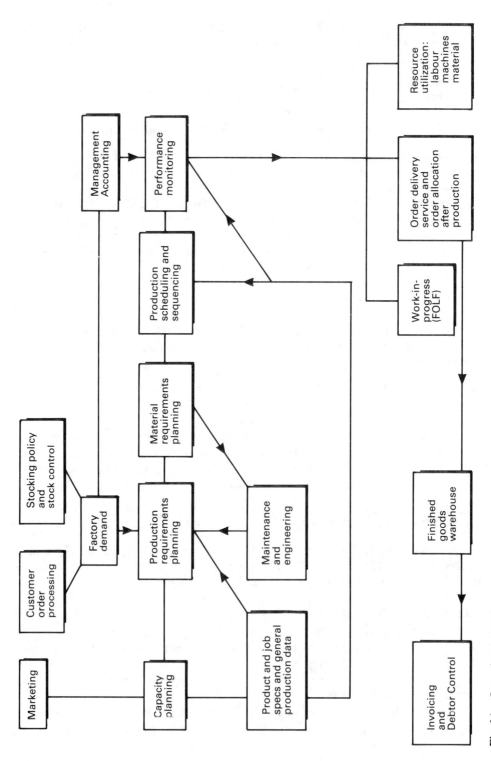

Fig. 6.1. Operational planning.

Hence an Operational Planning Department was set up with its first and major objectives to improve customer service while minimising working capital. A secondary objective, to reduce substantially the number of people involved in the separate functions which would now be amalgamated, was then established.

6.4.2 Operational Planning—a uniting function
We decided that 'Operational Planning' would sit at the heart of the company and be influential on all its parts. Past failings were to be redressed as follows:

(a) Reduce an over-extended product range.
(b) Improve forecasting and make it work more effectively in running the factory and utilizing working capital. Ensure that work content of products was better calculated.
(c) Re-vamp the supply allocation (i.e., orders coming off the works). This had always been dominated by the need to give absolute priority to Original Equipment customers. The replacement market suffered if splash demand came from an OE customer.
(d) Re-establish the balance between factory capacity and order intake. Combining a seasonal demand from numerous markets with stock replenishment orders, against badly defined factory capacities, never gave a good customer service and certainly did nothing for factory efficiency.
(e) Ensure that discipline in the factory was appropriate to achieving a production schedule precisely as it had been set down.
(f) Integrate associated systems so that company objectives were pursued, not just that departmental ones were pushed.

6.4.3 Operational Planning—a suitable organization
The following functions were put into the Operational Planning organization:

Forecasting
Order processing
Capacity Planning
Materials requirements planning
Production scheduling and sequencing
Performance monitoring/shop floor control
Stocking policy and inventory control
Order allocation
Marketing (but later taken away by Sales management, a retrograde step)
Purchasing.

It was appreciated that many systems as well as those contained in the above functions would impinge on Operational Planning as follows:

Management Accounting
Bonus/Wage Payment systems (pay roll)
Time and attendance reporting
Nominal Ledger

Sales Ledger
Purchase Ledger
Fixed Assets Ledger

6.4.4 Principles/objectives
We thought there should be the following features:

(a) Batch sizes and inprocess quantities (work-in-progress) should be minimal.
(b) The time between starting to make a product and its despatch should be as short as possible.
(c) Processing time in production should be minimal. Flexibility should be instituted wherever it was considered necessary and possible.
(d) Stock of all kinds should be reduced substantially.
(e) The planning techniques should include:

   – centralized materials requirement planning
   – bottleneck identification loading and control
   – planning of operations, both backwards and forwards from bottlenecks.

(f) Response times should be shortened considerably.
(g) Problems arising on the shop floor should be solved there.
(h) Production should flow quickly and output targets should be achieved day after day with no peaks and troughs.
(i) Actual customer orders and depot demand should drive the materials purchasing, planning and manufacturing systems.
(j) The simplest possible factory planning and control systems should be in use.
(k) Line balance was essential, i.e., the production line should be able to make all varieties of products ahead of it without creating major bottlenecks.
(l) Housekeeping on the shop floor should be very good, especially in moving orders in the sequence required. Rejects should be highlighted, their cause identified and corrective action taken immediately.
(m) Quality objectives should be clearly set out. Where these impinged on the achievement of production targets or machine efficiencies, suitable trade-offs based on cost data would be calculated.
(n) Plant loading should be uniform. Cycles should be established which synchronized production with demand. Systems scheduling should be used where production responded to demand, rather than having predetermined schedules.

We recognized that the key production personnel who would operate the scheduling and shop floor data collection elements of Operational Planning would be the first-line supervisors. Their training, attitudes and energy would be vital in a successful introduction of Operational Planning.

6.4.5 Benefits of Operational Planning
These appeared quickly and were as follows:

(a) A forum was established where appropriate data about capacity utilization, allocation of that capacity, service given to customers, stock levels and general production costs could be discussed usefully. A formal basis was brought into existence for taking decisions about increases or decreases in capacity for altering production manpower levels (up and down) and for regulating the need for overtime.
(b) There was a major reduction in finished goods and raw material stocks. Because of the intrinsic way the factory was scheduled and the outdated payment and work organization systems, work-in-progress was not at first reduced.
(c) Service levels improved especially stock service levels.
(d) Data on manpower levels, future output rates and general order demand became accurate enough to help in the debates with the unions. Trust was engendered because management appeared to know what it was doing.

6.4.6 Optimized production technique/zone control

Ferodo's response to the problems associated with a large variety of products and potentially 25 different production operations, was to institute 'zone control'. This had been formulated some twenty years before Operational Planning was brought into existence.

Production zones had been established in the factory, comprising five or so operations. One of the operations was designated a bottle neck or potential bottleneck activity. This operation was the one which was loaded/scheduled to its maximum capacity. The rest of the operations in the zone were not scheduled, but a broad measure of product work content in the order load and that potentially available was made.

If sufficient manpower was available to cover the total zone operations, there would be no reason why all orders committed to the zone should not be completed in a specified time. Zone control therefore was established as a mix of scheduling and controlling order flow. In some ways it was a precursor of the 'OPT' system which many companies have recently attempted to introduce. In the case of zone control all orders of a particular 'load' would receive a zone-completion date. Local supervision was enjoined to ensure that no order of the load was left in the zone once the date had been passed.

6.4.7 Desirable improvements in Operational Planning

It did not take too long for users to complain about Operational Planning and consider improvements. The debate about improvement followed these lines:

(a) It was essential that the systems linkages, especially between core elements of Operational Planning, related and interacted completely with each other. If one changed so should the rest, in proportion to the change. In other words we would really have liked the process we were tackling manually, but with some computer help, to be a computer-based model.
(b) It seemed possible to restructure the process so that a different, more relevant organization, might apply. Capacity planning, production requirements planning,

most stock control activities and despatch activities might form one central organizational unit, with production scheduling and sequencing, performance monitoring and customer order processing, another.
(c) Common database. Data needed to be established in a way which prevented duplication, either in the use of the information for the core systems elements or in those other functions closely associated with it.
(d) Minimum clerical effort was needed in either accessing the database or in updating it.
(e) Wherever possible information needed to be accessed on-line via VDUs with the facility to:

- roll screen display backwards and forwards
- move quickly between screens
- take a hard copy listing of screen data as required.

(f) Non-technical people should be capable of making ad hoc enquiries or specifying exception reports in user-designed formats, not those designed within the system. It was essential that systems data could not be corrupted, access to sensitive data was controlled, and simultaneous transaction standard processing times not be unduly affected.
(g) The system should provide a means by which all planning, production, stock control and purchasing functions were integrated by:

- using common databases within the manufacturing control system
- enhancing the possibility of transferring data automatically between the database and other data handling activities going on in offices—e.g., personal computers, telex and facsimile systems.

(h) As far as possible the various packages in use to handle sales ledgers, fixed assets, etc. should be retained.

6.5 SYSTEMS SPECIFICATION/STATEMENT OF USER REQUIREMENTS

Any major advance on the Operational Planning system obviously needed a very careful evaluation of what could be done and how. While MRP II appeared to be an answer, it seemed complicated. Its inherent complexity (especially when allied with Management Accounting) demanded that a 'statement of user requirement' was needed. Such a 'statement' sets out in precise detail the system design requirements of potential users.

Consultants will prepare such documents for around £60 000 for a medium-sized company at 1990 prices. We used a format prepared by consultants Coopers Lybrand in other T&N companies, but decided that the experience of 'doing it ourselves' would be invaluable in applying MRP II.

Much of the subsequent detail of this part of Chapter 6 was first recorded in our User Specification. We had the Operational Planning manual as a starting point and all the associated experience. Whether we could have done as well without this experience is debatable.

6.5.1 What the users wanted from MRP II

While our Operational Planning system resulted (largely) from a decision to amalgate disparate departments, which in essence were carrying out interrelated parts of an overall process, the extension into closed-loop MRP II was extremely complex by comparison. Even so, deferring the writing of the User Specification to systems specialists or consultants seemed to defeat the purpose of the specification. It was a 'User Specification'.

There may be a problem in getting line managers to play an appropriate role in establishing what systems they need and to record appropriate data. Spare time is never available, but not knowing what to do about MRP II is not an acceptable excuse for line management avoidance of the task.

6.5.2 MRP II characteristics

These are produced so that everyone will understand the nature of the proposal.

MRP II, or manufacturing resource planning, is a computerized system, based on a well-structured database, which will, by using lead times and relationships between production capacity, stock forecast and actual demand, determine a master production schedule. This will match demand with capacity in a way which will best serve customer service and general resource utilization.

The 'closed-loop' element or feedback mechanism tests the schedule and all other activities, so that the whole process can be re-run until an apparently optimum result is achieved.

In many ways, the best (and necessarily the most advanced MRP II systems) will provide a model of the *manual* manufacturing and inventory control process which could be used by managers to provide the best possible plan to run the total company process.

By itself, MRP II may not generate major cultural changes within an organization, but in generating a need for discipline and organizing resources to meet customer demand, it is essential.

The modules covered are:

Master Production Scheduling
Materials Requirements Planning
Shop-floor scheduling
Material purchasing
Performance measurement.

The whole is shown in Fig. 6.2. It will be seen that MRP II is used to:

— establish, or cause the establishment of, appropriate production capacities;
— recognize priorities to be made;
— ensure that the products prioritized are actually made accordingly;
— control stock levels and stock service;
— motivate raw materials and components purchases;
— help establish and monitor production performance.

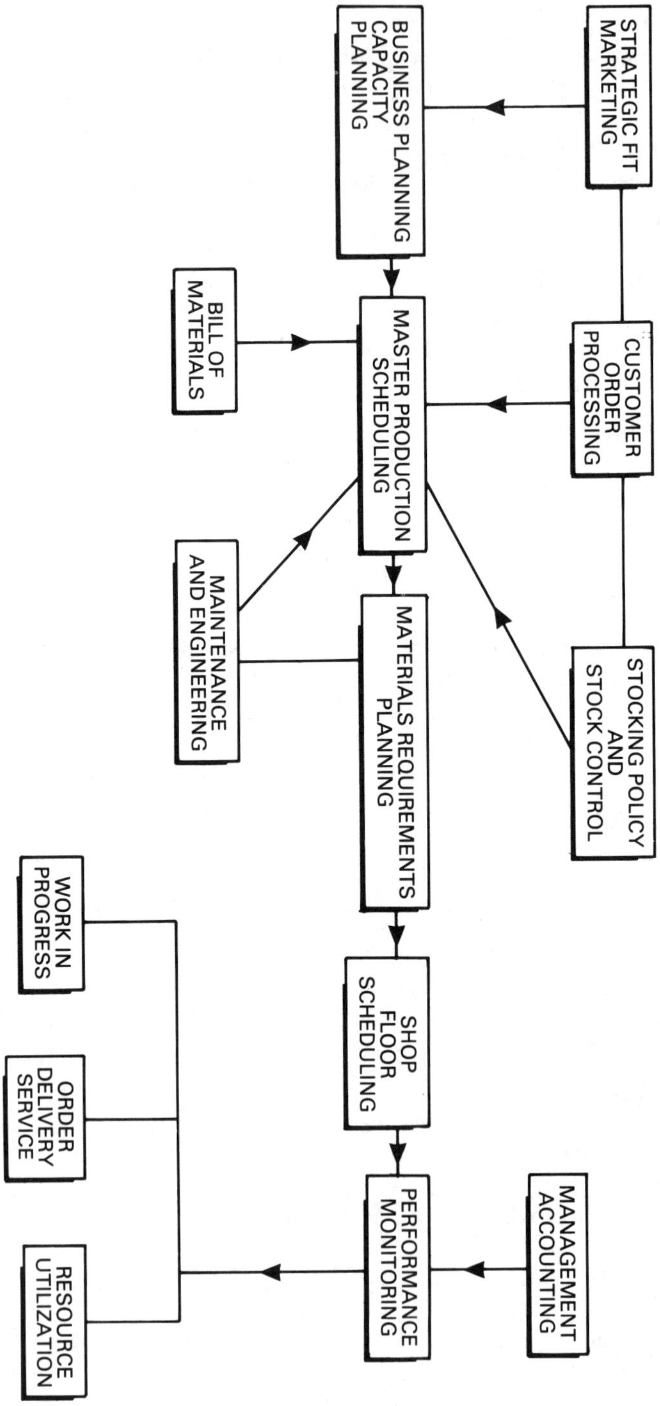

160 Systems and information technology [Ch. 6

Fig. 6.2. MRP 2—the activity.

6.5.3 Systems specification outline format
An outline format of our User Specification is given below.

(a) Key statistics—these will differ from organization to organization, but in our case constituted the following:

- Product markets
- Items on stock file
- Product groups
- Sales
- Sites
- Stock
    - value
    - items
    - location
    - targets/achievements

- Volumes:
    - customers
    - sales orders per month
    - average lines per order
    - despatches per month
    - invoices per month
    - purchases per month and average lines per order
    - supplies and products
        - raw materials
        - components
        - re-sale

- Stock transactions per month:
    - work-in-progress
    - finished goods stock
    - raw materials
    - consumables

- Manufacturing cost centres
- Units of plant
- Tools/jigs
    - operations per order
    - orders per month
    - average batch size
    - minimum batch size
    - profit and loss transactions per month
    - nominal ledger transactions per month
    - payroll monthly postings
    - plant register in units

- Information might be needed by:
  - product line
  - product group
  - product market.

(b) Current systems development—a brief description of each is needed with a record of any known failings, e.g.:
  - Lack of integration between individual systems
  - Too much reliance on informality and bypassing of existing computer systems
  - Continued use of old computer technology
  - Major elements of the system being badly defined
  - Mixed functional responsibilities.

(c) What planning/manufacturing/organizational/systems principles should apply?

(d) What new organization/s appears to be necessary and what new interfunctional relationships will be beneficial?

(e) What basic data is required? How will this be constructed onto the appropriate database?

(f) Systems design elements. For each element, the following will be needed:
  - definition of the element
  - objectives
  - data to be maintained
  - inputs
  - processing
  - outputs
  - interfaces.

(g) Other information outputs, e.g., Board or senior management requirements.

(h) Data collection—a separate review might be useful. Especially of shop-floor data collection.

(i) Details of software already available.

These nine divisions might be extended, but essentially the most detail will be required under (f) and the following may all need to be considered in the way outlined:

Capacity planning
Production requirements planning
Inventory control
Production load building and scheduling
Materials budgeting and purchase controls
Working capital.

6.6 THE INITIAL DEBATE

6.6.1 Block diagrams

In whatever way or form Operational Planning was constructed, it produced a complex set of interactive relationships. Even when the factory was separated into product lines and 'focused' there were still too many things that could go wrong, too many orders and too many product items to ensure that a simplified view of producing a master production schedule could be taken.

So, an important step in debating the need for MRP II and what it constitutes is to design block diagrams and relationships, which can be used to discuss why MRP II is important and essential.

We took Fig. 6.2 as a starting point and then produced block diagrams similar to those in Fig. 6.3, 6.4 and 6.5. To achieve a capacity plan which made sense, we knew that we had to define for each product range group:

- Data which will influence capacities:
  - delivery service required
  - seasonality
  - proposed stock levels
  - JIT
- Marketing plan/sales forecast
- Bill of materials
- Cost information
- Current capacity information and as an output
- Required capacity

For production requirement planning, we had to record constraints in production, the priorities established by sales, contribution, engineering problems, and whether tooling and raw materials would be available at the right time.

In each case we obviously needed maximum computer assistance, especially in being able to interrogate and relate to data in the capacity file, bills of material files, inventory and customer order files, etc.

Slowly we realized why we had to throw most of the old stand-alone systems away, re-vamp/re-create the company database and begin to established a system which would be able to operate within a closed loop.

6.6.2 Database
Most of the relevant information which would be needed for MRP II had already been collected during the establishment of the Operational Planning function. However, the importance of accuracy was perhaps not emphasized enough.

An integrated database was required for:

- Materials Requirements Planning
- Manufacturing Order Control
- Management Accounting
- Quality control/Statistical Process control
- Shop-floor data collection
- Capacity Requirements Planning
- Bonus Systems/Wages Payment
- The establishment of specifications for raw material and components bought in
- Maintenance planning and control.

164 Systems and information technology [Ch. 6

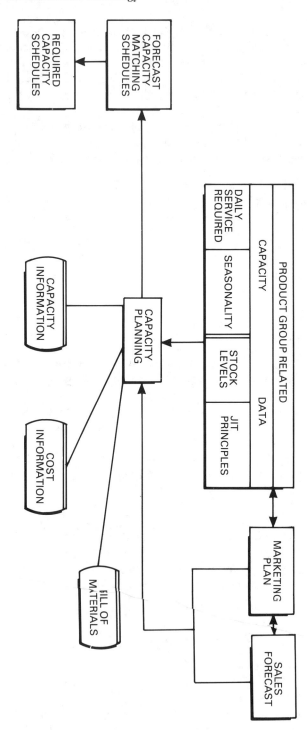

Fig. 6.3. Capacity planning systems/file design.

Sec. 6.6] The initial debate 165

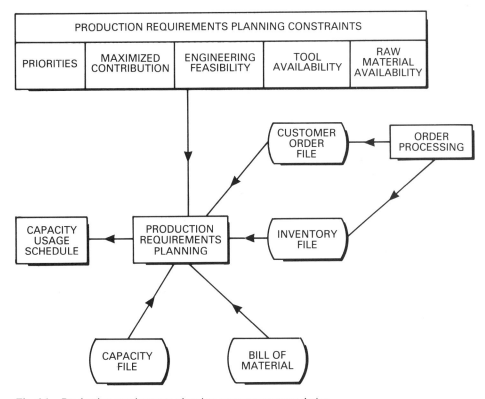

Fig. 6.4. Production requirements planning computer systems design.

The data needed was as follows:

- Making particulars
  - weight and cost of raw materials for 100 standard products
  - operations/activities needed to make the products
- Job rate list
- work study or other labour output values per activity
- Materials mix
- Manufacturing operation data
  - reference number of each product
  - unit of measure
  - material quality
  - size
  - material yield
  - tools required

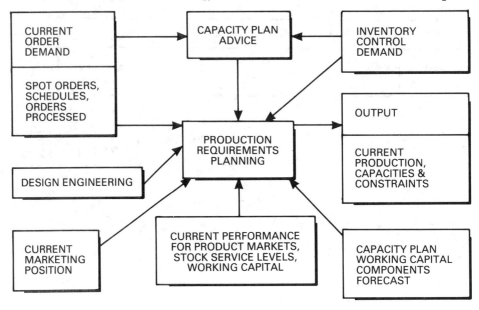

Fig. 6.5. Production requirements planning as a managerial activity.

- Operational information

    - code numbers
    - operations
    - standard minutes per 100 products per operation
    - standard operating cost
    - quality control information

- Planning data

    - standard lead times for manufacture
    - inspection time
    - bottlenecks, or limiting factors
    - current work load/WIP

- Stock control/re-ordering

    - current stock values
    - stock targets
    - batch sizes
    - type of stock items (G., E., or T.)†

†The Ferodo equivalent of standard 'ABC' classification.

- Ordering policy—whether, for example, over-ordering is possible, leading to over-stocking to keep the factory operating at agreed levels.
- Purchase routines
  - policy
  - safety stock
  - lead times
  - supplier profiles
  - make/buy decision rule

- Machines/operations tools
  - capacities/output rate
  - standard capacities/current capacities being achieved
  - tools (numbers, types, etc.)

- File structure—data files already existed. Reluctantly and regretfully, these were thrown away and new structures put in their place, compatible with the MRP II package we were to use.

The database has to accept:

- Inputs
  - changes to information, e.g. new additions to the product range
- Processing
  - mainly that associated with the control and validation of file maintenance transactions
- Output
  - error reports on missing or corrupted data parts list
  - operations/activity/routing reports including cost centre analysis/manning/payment levels, etc.

Interfaces
The database provides information for multi-systems use, viz:
  Materials Requirements Planning
  Management Accounting
  Manufacturing order control/factory load building
  Manufacturing database
  Inventory Recording and Control
  Purchase order management
  Shop-floor data collection.

6.6.3 Design of the core elements
Once the database had been specified in detail the next step was to designate the key elements and record the user requirements in each case. Production Requirements Planning has been chosen as an example:

Production Requirements Planning

We understood Production Requirements Planning to be the short-term application of the general rules on outputs specified in Capacity Planning, i.e., in matching capacity with demand.

(a) Monthly review

Each month the Operational Planning committee met to review:
- current/immediate output being achieved—in pieces and standard hours, comparing it with previous targets and profit plan per product group
- bottlenecks and other capacity problems
- current stock levels versus targets by product market
- current stock service levels by product group and market including GET[†] service levels.

It then determined the needed:

- increases/decreases in output rates
- increases/decreases in finished goods stock
- increases/perhaps decreases in stock service.

So, in essence, the committee set the future activity level of the factory in the light of current output level and stock. Once set, there was a commitment on the part of the factory to achieve what had been set.

Current and future sales were also discussed, especially where under-utilization of installed and, more particularly, planned capacity usage was taking place.

(b) Data to be maintained

This was as follows:

- An updated forecast, probably using exponential smoothing
- Actual order input expressed in
  - pieces
  - standard hours
  - material types
- Input—from individual product markets and the inventory control system, all compared with profit plan forecasts
- Revenue/contribution/ to be earned from current input compared with plan
- Current stock levels—GETO, etc.
- Stock service levels
- Profit plan requirements by product market
- Output achieved in the last 3 months compared with that required
- Orders already committed (forecast load)
- Current capacity usage
- Delayed orders

[†] The Ferodo equivalent of standard 'ABC' classification.

- by product market
- in pieces, revenue and contribution
- Current tooling situation
  - availability
  - downtime due to lack of tools
  - machine efficiencies.

(c) Processing

We needed a statement which showed target capacity utilization in each product market which would:
- satisfy current demand within required lead times
- even out production capacity usage
- achieve stock targets and targeted levels of stock service
- take account of production/material/perhaps cash constraints.

(d) Output

The main output we needed was a set of capacity usages in volume standard hours, work content and contribution, in the detail which would allow:

- materials management personnel to proceed to load build and schedule relevant product lines
- local bottlenecks/limiting factors to have been recognized within the process calculation

A working capital forecast was also required.

(e) Interfaces

The Production Requirements Planning needed to be interactive with and utilize information from:

Sales order processing/Contract scheduling routines
Product master database
Planning rules file/Capacities and constraints file
Inventory Control/Stock file
Capacity Planning
Purchasing order management system.

(f) Stock and cash forecasting model

As we considered stock levels and cash generation to be very important, we stipulated that we needed a cash/working capital forecasting model, which could be generated from:

- Future sales revenue—by product group
                               market
                               major customer
- A forecast which needed to take account of current terms of trade and debtor days
- Forecast stock values—for raw materials

work-in-progress
finished goods stock
- Creditors
- Extraneous payments and receipts.

6.6.4 Other elements in MRP II
We also needed to record, and re-specify where necessary, other systems as follows:

(a) **Inventory Control/Stock Recording**—mainly the Cascade stock control system which we had devised. Accurate stock records are essential in MRP II. Normally, owing to a whole range of problems, from cut-off failings to misrecording, they are rarely accurate.
(b) **Material budgeting and purchase control**—although this had been an integral part of Materials Management/Operational Planning for many years, relationship data needs and processing requirements all needed to be specified.
(c) **Order processing**—each product market needed a slightly different system, with order processing speed the linking requirement.
(d) **Management Accounting**—considered further, later in this chapter.
(e) **Other related systems**—e.g. Purchase Ledger
Sales Ledger
Nominal Ledger
Sales Analysis
Pricing
Invoicing
Payroll.

6.7 IMPLEMENTATION OF MRP II

6.7.1 The MRP II introduction plan
Does the implementation of an MRP II system differ from that of any other system? Well, it is bigger and it will have more impact on the company than any other system. It will certainly need more understanding than anything which has probably gone before. It could be a complete and very costly failure; or it might be a watershed, leading to a fundamentally different approach to running the company.

Hence the need for training.

There have been many major books written on the implementation of MRP II[†]. We used MML Ltd (Refuge House, 2–4 Henry St, Bath BA1 1JT) to help in our training. It is difficult in this respect not to use consultants. Training/education, it is said, can make all the difference between success and failure.

So our initial plan and then experience in introducing MRP II was as follows:

[†] E. G. Thomas and F. Wallace. *MRP II. Making it happen.* Oliver Wright Publications. 1985.

Sec. 6.7] Implementation of MRP 2 171

(a) Discover what MRP II is, what it can do and what its potential failings are. Talk to users. Do not accept anything that consultants say. Implementation, we discovered from others, is difficult, time-consuming and costly. If senior management is not prepared to give a considerable proportion of their time in seeing the project through, it will not work.
(b) It is very costly. It is practically impossible to develop one's own system. It is costly to buy a package, to train, to implement, finally to run the system. Potential costs should be worked out very carefully.
(c) If a package is to be bought, then look at several. Discuss their application with users. Halfway through our MSA package application, MSA decided to withdraw some support for package development.
(d) Get commitment. Even before extensive training begins, everyone should be committed to MRP II. If the Chief Executive believes that it is just another computer system put forward by those mercenaries in the Management Services department, it is better to forget the whole project.
(e) Write out a User Specification. Consultants suggest that documenting the current system is time-wasting and unnecessary. We found this not to be so. We identified what we had and what we still needed to install MRP II. We considered which computer system we would continue and not integrate with the main MRP II system. We recorded carefully the available data.
(f) Establish a 'game plan' of performance which MRP II should help achieve, e.g., return on capital employed, operating profit/sales, working capital/sales. Again re-assess benefits against the known costs. Is it worth it? Can we do without it?
(g) Set up a team to carry out the implementation. By far the most important member will be the project leader. He must understand MRP 2 thoroughly and be fully committed to it. Democracy can be a feeble thing in this situation.
(h) Set an implementation schedule. All users of MRP 2 state that achieving a 95% accuracy in stock records and a much higher accuracy level still in the database is a vital starting point. We, also, found that this was the case. There is no avoiding the checking and re-checking of data on products, materials, routines, timings, capacities, which apparently have been well known for a long time, but in reality are inaccurate. Stock recording, too, needs to be accurate. It seldom is in real life—here is a chance to make it so.
(i) Establish the order of march on the way the package is to be implemented and what is to be left out. In our case, we did not consider the costing module, for example.
(j) *Train, train, train,*—in understanding MRP II, in understanding the business it is supposed to plan and control, in setting and monitoring objectives, when it is required to re-run the whole system.
  DP/systems analysts will probably need separate training from general/line management. We took training 'down' to first-line supervision and shop stewards.
(k) Achieving a relevant Master Production Schedule should probably be the first main goal. It might be introduced as a 'Big Bang', not a parallel run or pilot study.
(l) Continuously audit progress. Success and failure needs to be monitored every week and every month.

6.7.2 Some general points
Our general experience suggested that:

(a) MRP is a closed-loop system. Everyone should know this and use the system accordingly. MRP I is still discussed in some text books. It is the designation given to a Material Requirements Planning system in which order inputs are calculated along with raw material requirements. These are related to current production capacities and a master production schedule is drawn up.
(b) MRP II takes the process a stage further and includes most other resources needed to make effectively a model of the company's manufacturing activity. It is a major system covering all activities from sales order processing to despatch. It is not an adjunct to the general production system activities. It should be the means by which these are planned and driven.
(c) The complexity of the system makes it difficult to alter or amend. Managers must take it as it comes. There can be few second thoughts which will not be very costly.
(d) Application is lengthy and costly in management's and system analysts' time. We began to see some daylight nearly two years after we started. The missionary spirit can die during this time.
(e) By far the most disconcerting cost was that of actually running the system.
(f) The system tends to work best when there is more installed productive capacity than is going to be needed in the next planning period. Having actual or potential bottlenecks increases the complexity of the system. Building up the number of direct operatives beyond those needed immediately is also expensive, but it will help MRP II to function properly.
(g) MRP II gives managers power to alter plans and schedules on line. On-line sequencing should become routine. However they should know that what they do influences all associated factors as well as their own.
(h) Rough-cut capacity planning enables the master production schedule to be changed in an interactive way.

6.7.3 Conclusion
There is a tendency for managers to look overwhelmed at the complexity of MRP II and the difficulty of introducing it. The 'How do you eat an elephant?' question arises, along with its answer—'a bite at a time'. The 'bite at a time' is a thoroughly worked out plan, rigorously monitored.

Consultants and writers may reel off the 'Ten Commandments' of introducing MRP II, but the best advice is to consider the associated software as only 10% of the problem: the rest is training, testing, using and achieving objective. Accurate data is essential when achieving a good master production schedule.

The number of Class A MRP II users is still limited. It is possible that, again, we have a technique which is too demanding on the internal discipline, human relationships and management commitment, to succeed.

While MRP II is important in its own right, it will never work effectively if all the other parts of the manufacturing strategy are not in place as well

6.8 MANAGEMENT ACCOUNTING AND BUSINESS PLANNING

6.8.1 Introduction

A fairly concerted view by advocates of advanced production technology is that the traditional methods of management accounting should no longer apply. Many of the supposed benefits of advanced manufacturing technology, it is said, are intangible and cannot be quantified within a standard/traditional investment appraisal. It is difficult, they go on, to project benefits over a long period. The company-wide benefits are ignored by management accountants.

Precisely these words were used to persuade companies to buy first-generation computers. The results were mainly poor.

Robert Kaplan is quoted by David Walker in the *Financial Times* of 18 September 1989 as saying,[†] 'Today's management accounting information, driven by the procedures and cycles of the organisation financial reporting system, is too late, too aggregated, too distorted for managers' planning and controlling decisions'. The main contentions expressed were that managers receive misleading information from short-term financial reporting and so take wrong decisions about crucial issues such as product mix and capital investment.

This kind of comment worried us. The management planning procedures certainly seemed to need improvement; so we decided to review the whole activity.

We again wrote a 'user specification' where principles, policies, requirements and possible systems changes were listed. Despite the improvements that had been made in the last decade, there were still gaping holes in the management accounting process.

The system was still not an integral part of the company's business information systems. It helped, but could have been much better in improving the operating performance of the factory. Again, it helped but needed to be improved in maximizing contribution.

Despite constant updating and correction, standards of performance in the database were incorrect to some degree, and this helped to give rise to erroneous and misleading variances. Structuring to give information by hierarchy had been improved but was far from perfect.

Perhaps the greatest disappointment was that the Management Accounting system was still not all-pervading among managers and supervisors, with them constantly leaning on it to make decisions.

All these aspects were re-stated as part of the 'user specification'. Others were:[‡]

(a) Timing was important and information should be produced which would help to institute corrective action when things started to go wrong. Whether this was hourly, daily, weekly or monthly did not really matter.
(b) All reporting times needed to be shortened.

[†] *Relevance lost in the rise and fall of Management Accounting.* H. Thomas Johnson and Robert S. Kaplan. Harvard Business School. 1987.
[‡] Taken from the Ferodo User Specification.

(c) Contact between Management Accounts staff and users of the information was often desultory and unproductive. Some accountants had bureaucratic inertia to the extent of debating results with line managers even when they were obviously wrong.
(d) We knew however that there were wide implications in the use or misuse of Management Accounting, for example in the top-down planning and use of company models.

While Kaplan recommends something of a revolution, the Association of Cost and Industrial Management Accountants suggests that a more gradualist approach is needed.[†] Overhead allocation, the curse of absorption costing, is one problem which will go on bedevilling production units while costs are allocated on a production hours basis or some other not very useful factor. Kaplan is very firm about not paying too much attention to direct costs, as these are too small to be really significant. This seems sensible—until direct labour is determined as having a key role in improving machine utilization, efficiency and material yield, and gaining added value. Then it becomes very important.

We considered 'activity costing', where all the elements of cost, beginning with raw material suppliers and down to despatching, are taken into account; but the whole process is extremely complicated. Where product prices are largely predetermined by the market, it appears to produce only minor benefits.

6.8.2 Line management involvement in Management Accounting development

'Management must determine its own requirements' has been a fairly constant cry from those who design systems and, in particular, management information systems. In practice, this is not normally very fruitful. The comment assumes that line management (in this case) is both fully conversant with Management Accounting and knows, completely, what information will aid the line management process. This is indeed rare, and managers need to be coaxed and, above all coached into accepting information weaknesses and what needs to be done to put them right. With more help from the systems designers than would conventionally be forthcoming, this list was compiled:

(a) General
The Management Accounting system must be capable of producing information for a variety of decision-making activities. It should be multi-faceted, changing the basis of the information easily and rapidly. For example, for pricing purposes, the variable cost of manufacturing and selling is necessary. In comparing two alternative manufacturing methods, the variable product cost alone might be used.

(b) Variances from plan
Speedy reconciliation of achieved results against plan, with effective variance analysis which enables needed responses to be introduced as quickly as possible.

[†] *Management Accounting—Evolution not Revolution.* M. Bromwich and A Bhimani. CIMA. 1989.

(c) Budgets

It would be useful, if possible, to agree what kind of operating profit on a business, product market, site and company as a whole was necessary. Then one ought to have the ability to model activities, their revenues, costs and contributions and to determine, after due analysis, how the anticipated profit could be achieved.

While setting budgets should be done on a participative basis, the procedure must be regarded as dynamic since there will often be a need to re-deploy resources during the financial period for which the budget is being set. The best use of scarce resources should follow a bidding system based on potential results.

Budgets should be agreed only after well-conceived procedures for their scrutiny and evaluation have been followed.

(d) Fixed capacity budgeting

The system should provide an ability to start with installed production capacity, to determine what can be made at what cost and then to review marketing strategy to assess how far sales revenue/contribution can be gained which will maximize utilization.

(e) Exception reporting

Senior management is often bogged down with too much, rather than too little, information. As a result, key data is often overlooked and not given the importance it deserves. Exception reporting of key events should be used much more than it is now. These factors alone could go a long way towards improving even the current systems.

(f) Cost comparison

Accessing cost data via desk-top terminals seems a crucial part of examining elements of the company's business and determining how far cost reduction needs to be taken to achieve new business. So, as well as providing product costs and the standard performance reporting procedures, we must be able to ask the 'What if we did this?' type of question and get a reasonably accurate answer.

(g) Product range

Despite major reductions in the product range, it still seems too big. We need Management Accounting information which either justifies the current range or enables us to reduce it.

(h) Standards and standard cost

We appreciate that direct standard costs, accurately determined and understood by all local management, needs to be the core element in the costing system.

(i) Contribution and absorption costing

Local management is aware of the differences between contribution and absorption costing and the surrounding debates. Emphasis on contribution linked to fixed capacity budgeting seems to give the benefits of both systems. The reconciliation

(j) Speed
Monthly accounts often report on activities which took place anything up to six weeks previously. Using such accounts to improve performance can be counter productive. Each Monday, a satisfactory sales and product variance operating statement reporting on the previous week's activities is required, so that a significant response can be made to continuing problems.

(k) Performance monitoring
The annual plan, once made, should be achieved. There are few acceptable excuses for not doing so. Most performance monitoring, therefore, should be directed towards showing local management how well they are achieving their part of the overall plan and indicating where performance needs to be improved. Variances from plan should be recorded, along with reasons why there are such variances. The causes for failure should be reported as unambiguously as possible.

The following major areas of potential concern need to be monitored:
(i) Actual revenue earned compared with that planned—reason for any variance to be explained as changes in volume, price, contribution, mix; by market, product group, major customer, country or major product items.
(ii) Actual factory cost compared with standard—variations in production volume, labour and machine efficiency, labour rate, material yield, variable overhead, other variable cost, fixed cost, explained.
(iii) Trading expenses variances from plan.

(l) Standards
Establishing operational standards which can stand up to the real world of manufacturing diverse products of a fairly large product range is a key element in profit planning and control of production performance. A 'normal standard' will be an agreed level of performance which if achieved will come nearest to achieving organizational objectives, when all operational constraints have been taken into account.

Standards should accept recent operational performance and subsequently use it so that operational variances are minimized or easily explained.

The standard cost of output should be the basis of performance reporting in the factory and actual costs compared against it. 'Standard hours' should be the basis of establishing and monitoring labour productivity.

Annual reviews of standards should be very serious occasions when past, present and future performances are debated, conclusions drawn about inadequacies in planning, resource utilization and management competence. Standards should then be re-set, which means that local management is stretched, especially if there has been capital investment.

Some accountants argue that amendments should be made to standards throughout the year. This is nonsense. It makes comparing the profit plan and actual result very difficult, if not impossible.

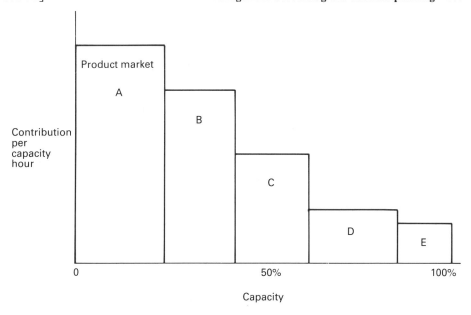

Fig. 6.6. Histogram showing contribution earnings of a production line.

(m) Types of cost

The most important costs in the production and distribution areas are those which vary with the activity being carried out. One of the most useful calculations is to determine the net sales revenue being gained for products and the variable or direct or marginal costs being incurred in their manufacture and distribution. The difference between the two is 'contribution', as it 'contributes' to overheads and profit.

Contribution can be used in a variety of ways. If production capacity is being fully used, for example, it is likely that histograms of the kind shown in Fig. 6.6 can be produced. Some parts of the capacity are achieving good results, others are not.

Maximization of contribution should be debated constantly between production and sales personnel. Orders should be accepted, mainly on the basis of filling current capacity, while maximizing contribution. Salesmen are not usually adept at changing the sales mix quickly. Production personnel should insist that they quickly improve on their ability. For production managers then,
  direct labour,
  direct material,
  variable overhead,
  energy,
  and any other variable cost,
must be controlled closely and, if at all possible, reduced, There are two ways of improving contribution—reducing variable cost and improving net sales revenue. Both are equally important.

(n) Limiting factor analysis

The principles of optimal production technology (OPT) enjoin production engineers to balance flow not capacity. To treat all operations and processes as being equal is to give some an unreal importance. The cost of a stoppage at a bottleneck is infinitely more important than at a point where there is over-capacity.

For many years, management accountants have advocated the use of limiting analysis, and indeed Ferodo's Zone Control system was based on this appreciation. The cost accountant needs to emphasize more and more the loss of contribution at a bottleneck. Equally, he should help to maximize contribution by ensuring that priority is given at the bottleneck to those products which give the most contribution per bottleneck hour.

(o) Production Operating statement

The basis for portraying the variances from plan should remain operating statements. A common data collection and processing system will be needed to ensure that there is consistency and integration of reporting procedures and data used within them. As far as possible, manual intervention in data collection should be minimized, but 'shop-floor data collection' with operatives reporting directly into the system via data terminals may not be necessary.

Data collection should concentrate on obtaining an accurate record, shift by shift, of:

- production achieved
- direct labour hours utilized
- material utilization and rejects
- variable overhead expenditure.

Additionally, non-qualifiable reasons for variance need to be collected.

(p) Relationship with MRP II

The requirements of local management should drive the introduction of, or improvements to, MRP II and Management Accounting.

6.8.3 Elements of the system

(a) Information databases

Some of the more important information needed in the database was recorded in the first part of this section dealing with MRP II. The databases should be used for both MRP II and Management Accounting. Its relevance to Management Accounting should be in:

(i) helping to establish relevant standards and budgets and in changing them at the end of the financial year;
(ii) calculating individual product cost;
(iii) directing attention to the products, activities or functions where improvement is (necessary;
(iv) helping to monitor performance in production, associated factory activities, sales/markets, and S, D, and A areas.

(b) Activity rates
Activity rates should be established for operation/product items or groups. Three rates are required:

(i) The technical maximum i.e. the maximum limit at which plant is designed to operate;
(ii) Standard performance—normally the technical maximum, less all policy decisions concerned with such things as 'clean downs' at the end of the shift, tea breaks, etc.;
(iii) current performance—whatever is the historic average being achieved.

Using these three activity rates, local management should be constantly reminded of the utilization (and especially the under-utilization) of machines which is taking place. This should be valuable in setting rate standards.

(c) Standard variable cost
This is a cost based on the standard weight and cost of materials, standard hours of production and associated cost, plus other variable cost, all added together. It should form the basis for crediting a cost centre with the values of output as well as being a part of a standard variable product cost.

(d) Variances
Calculation of the following variances are needed:
Sales volume
Sales price
Product mix/market mix
Production volume
Production efficiency
Rate variance
Material variance
Finished goods stock and WIP movements.

(e) Methods of reporting
The term 'operating statement' covers a series of documents and reports which record the operating performance of the company and particularly the production function where most costs are incurred.

Two types of statement are needed. The first should give a monetary evaluation of performance compared with either standard or budget. The statements will facilitate management by exception and relate to local managers' objectives. The second set should be non-monetary.

The main statements need to be:

(i) Company/Line/product state (Fig 6.7).
(ii) Business reports—any contention between reporting business, site and product line performance should be reconciled on this form (Fig 6.8).
(iii) Daily reports—these reports reduce the time gap between actual performance and report to an absolute minimum. (See: First-line management reporting, in Chapter 5.)

(iv) Weekly operating statements (Fig. 6.9). These report on the financial situation evaluated from the daily report.
(v) Product line operating statement.
(vi) Business information schedule—reconciles plan with actual results by business (Fig. 6.10).

	ACTUAL	PLAN
SALES		
VARIABLE COSTS Material Labour Overheads		
TOTAL VARIABLE COSTS		
CONTRIBUTION Fixed overheads Fixed overheads stock move Capitalized fixed overheads		
GROSS MARGIN		
TRADING EXPENSES Sales Carriage & distribution Selling expenses Administration R&D Other income (-) expenditure TOTAL TRADING EXPENSES		
OPERATING PROFIT Interest Technical fees Royalties received Dividends received Exceptional charges		
PROFIT BEFORE TAX		

Fig. 6.7. Profit achievement statement.

Month:

(1) Budgeted sales revenue less (2) Budgeted direct/ manufacturing cost					

(3) Budgeted gross contribution
INCREASE/DECREASE IN CONTRIBUTION DUE TO SALES PERFORMANCE
(4) Sales volume
(5) Sales Price
(6) Product mix
(7) Market mix
INCREASE/DECREASE IN CONTRIBUTION DUE TO MANUFACTURING PERFORMANCE
(8) Direct labour variance
    efficiency
    rate
(9) Material variances
    usage
    price
(10) Variable O/H variances
    efficiency
    spending					

(11) Contribution before adjusting for (12) Finished stock & WIP movement (variable)					
ACTUAL CONTRIBUTION					
(13) Manufacturing fixed cost budgeted plus/ less variances					
GROSS MARGIN					
(14) SD&A costs budgeted					
(15) OPERATING PROFIT					

Fig. 6.8. Business Reporting reconciliation.

Budget standard hours	Planned standard hours	Actual output	Labour utilization		Material utilization		Variances				Stage Contribution hour rate Date
			Standard	Actual	Standard	Actual	Efficiency	Labour rate	Material usage	Other	

Fig 6.9. Weekly operating statement.

Sec. 6.8] Management accounting and business planning 183

Product Customer groups	Budget for month					Actual					Variance			
	Net sales revenue	Standard variable costs	Contribution	P/V ratio		Net sales revenue	Standard variable costs	Contribution	P/V ratio		Price	Mix	Volume	production

Fig 6.10. Business information schedule.

These operating statements will inform local management as to whether they have gained or lost money in their part of the company compared with standard and plan. They should be capable of relating to each other. The reasons for variances should be obvious, or quickly become so.

Other reports might be as follows:

Customer contribution
Product contribution
Cash/working capital operating statements
Measurement of added value per product group
                         major product
                         machine hour
                         direct worker
Wages paid and output achieved (to help determine the degree of non-productive time).

Planned sales budgets should be measured against:

Current factory capacity
Potential contribution to be earned
Stock and other resources needed to support sales
Economic considerations - competition, prices, inflation.

Planned production budgets should be viewed against:

Standards in the database
Past performance
Need to improve performance.

Planned administration budgets should be reviewed against:

Service given
People employed, with a suitable relationship established between costs incurred and money spent, e.g.
− orders processed per £
− records processed per £
− documents processed per £.

Sales, distribution and administration costs.
Research and development costs.
Capital expenditure proposals.

The further factors which senior management should review will include:
Working capital needs
Product mix changes
Limiting factor analysis and plans
Sales prices
Exchange rate movements
Cost inflation including potential changes in:

- labour rates
- raw material prices
- other bought-in items, e.g. electricity.

A comparison should be established between the required revenue and cost performance needed to achieve overall objectives and that which senior managers believe they can achieve.

A rather painful procedure might then ensue, when a rigorous debate on how gaps between company need and senior managers' offer might be closed takes place.

6.8.4 Top-down planning

This process has been mentioned several times as a method which gives results much superior to 'bottom-up budgeting'. Top-down planning is based on the Board of Directors setting broad objectives for the company, its sites and businesses. These should follow a review of product markets, past and present results, the requirements of the City, etc.

The sales forecast should also be used as a secondary piece of information. If the match between forecast and installed capacity is too great, the sales/marketing people need to go out and get more prospective business—somehow.

The strategic fit between product markets and the company must also be a potent factor in setting objectives.

The primary objectives should be:

- Net margin—the operating profit/sales ratio
- Gross margin
- Return on capital employed per:
  - business
  - product line
  - site
  - company.

Senior managers will then need to review and agree the following:

- The Revenue Plan
- Direct costs

  - labour
  - material
  - energy
  - variable overheads

- Fixed factory costs.

6.9 DEVELOPMENT OF A MANAGEMENT ACCOUNTING SYSTEM— A SUMMARY

Much of the foregoing discussion relates to the setting up of a User Specification designed to produce an improved Management Accounting system. While some of the basic accounting procedures have been left out, line management's needs have been stipulated. These are by far the most important aspect of developing a new system. Debates about whether activity costing/absorption or contribution are needed seem unimportant compared with establishing an understandable, if not entirely simple, set of procedures which line managers can use to do their jobs better. The development procedure can be summarized as follows.

(a) Costing principles
- Debate the costing principles and whether these apply or not.
- Review and record systems objectives.
- Establish methods to be used in determining:
  - production costs
  - product costs
  - variances from budget.

Review all elements of cost:
- direct labour
- direct materials
- variable overheads
- cost rates
- price level bases
- cost comparisons
- reviews of standards.

(b) Establishing plans and budgets
- Determine how top-down planning can be made to work, including setting broad-based objectives for:
  - businesses
  - product groups
  - resource utilization
  - return on capital employed
  - return on sales
  - turnover/capital employed
  - turnover growth
  - cost level improvements.
- The top-down plan should use agreed assumptions on:
  - market/economic conditions
  - potential turnover by product group
  - cost inflation
  - etc.

(c) Reporting procedures
Establish a reporting procedure as per Fig. 6.11. The information should form a reporting pyramid with each subsequent level receiving a summarized version of what

the previous level has used. Information produced should relate directly to local management's objectives. (All assuming that organizational hierarchies are still in place.)

People concerned	Data required	Time period
Board and directors		
Site senior managers		
Production managers Other managers		
First-line supervisors		

Fig. 6.11. Information hierarchy.

6.10 RECORDING DIAGRAMS

Business planning, Management Accounting, performance reporting have a tendency to frighten line management who do not have an accounting background. We thought that understanding 'Finance' was so important that we introduced a finance/business planning course which we gave to all management, supervisors and shop stewards.

The combination of computer use and business planning serves only to extend puzzlement. So it is useful to record, simply, the Management Accounting procedures. Figs 6.12, 6.13 and 6.14 show how product costing, material utilization and operating statement activities are carried out by computer.

The key factor is the PMDB, the product master database. If this in any way either is misconstructed or contains invalid or obsolete data, the whole business planning/Management Accounting procedure can go badly wrong.

6.11 POTENTIAL DEFECTS IN MANAGEMENT ACCOUNTING

What are the major reasons why Management Accounting and associated business planning fall into disrepute? These are some of the major problems:

(a) Standards are inaccurate and not kept up to date from one year to another. They are often the cause of fictitious production variances which cause major headaches.
(b) The system is directed away from production people who need it most, towards producing financial data which can integrate with financial accounts, which will then form the core of monthly reporting.
(c) Cohesion in reporting is missing. Though considerable data may be produced each week and month, its relevance, and especially its relationship to objectives, may not be obvious.
(d) Information may not be produced on-line, or if off-line timely.

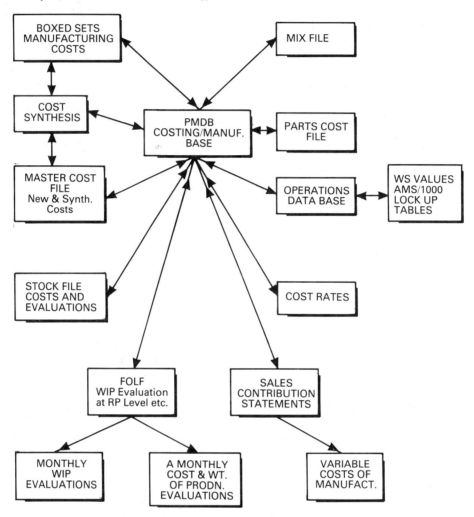

Fig. 6.12. Management Accounting—product costing. PMDB: product master database: contains manufacturing information (operations, etc.) for all manufactured products and work study (WS) values (AMS/1000) for direct operations. Accurate costs are calculated by referencing Mix File, Parts Cost File, Ops. DB and Cost Rates File, and are used for Work-in-Progress Evaluation, Stock Evaluation, Cost of Sales, Contribution and Variable Costs of Manufacture.

(e) Reporting times generally may be over-long. The shortest possible time gap between action and a report on it at appropriate management level, should be the aim.

(f) Contact between Management Accounting staff and users of their information could be minimal. The highest possible profile is needed by Management Accounting people, in seeking out how their function can better support production management.

Potential defects in management accounting

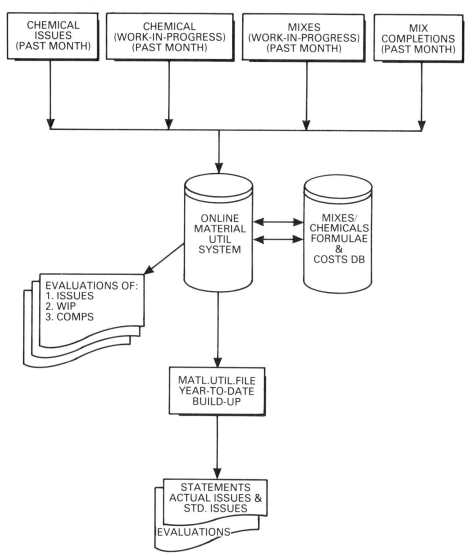

Fig. 6.13. Management Accounting—material utilization.

(g) Accurate reporting, especially via the work-in-progress file, and work booking may be missing. 'Cut-offs' could be a problem, with one set of data being reported as allocated to one time period, when the action being reported occurred in a different one.

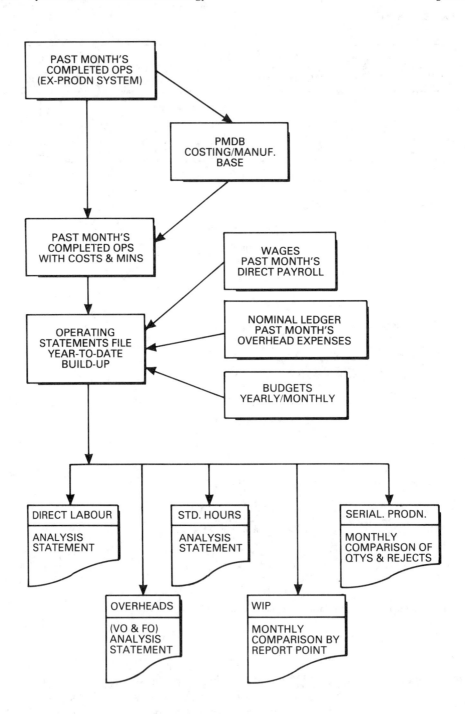

Fig. 6.14. Management Accounting—operating statements.

6.12 ABC—IS THIS AN ANSWER?

ABC, or activity-based costing, is being put forward as one of the most powerful financial tools that business and production managers can use. Its proponents suggest that it can lay down a cost accounting infrastructure which all world-class manufacturers need. They go further and state that dividing costs into direct, indirect and fixed, such as direct labour and direct materials, is far too crude to enable a production manager to take appropriate action. Such cost divisions and applications do not, for example, show the quality, which can be one of the key factors—perhaps even the absolute key factor—in ensuring the survival of the business.

As the fixed or overhead cost has grown in proportion to direct costs such as labour, it has become increasingly important to make sure that fixed cost is justified.

The fundamental change which ABC brings about is an identification of 'cost drivers' in the company—those elements of cost which really make or break the organization and its profitability. Quality has already been stated. Others could be planning (MRP II), or operating just-in-time, or product design, or production engineering.

The traditional cost allocation, so dominant in absorption costing, is abandoned and the cost drivers take over.

For example, it is likely that a low volume product might emerge from a traditional absorption-costing cost allocation, with a very low standard cost product cost (mainly because fixed cost has been allocated on a volume basis). However, it is more likely that a low volume item will consume, proportionally, more fixed cost than a high volume one. So the traditional product costing method can distort considerably any review of products and how profitable they are.

ABC should enable the allocation of cost drivers in a much more rational and useful way. It may be possible that part of production engineering is being used to support a small section of the product range. In other words, it is being misused, or could, with some careful analysis, be reduced.

ABC gives a view of costs and their benefits, therefore, which other methods of costing do not give. There should be a direct and important relationship established between what fixed costs are being incurred and whether they really are worthwhile in achieving company profit. This alone should motivate senior management in establishing ABC as a supporting element to a traditional costing system or indeed, with a contribution method, ensure that fixed costs are apportioned logically.

CONTENTION

(1) Anyone involved with developing systems or advocating the use of IT must have a strong business background. The failure of computers to be any more than shufflers of not very accurate information largely stems from too little involvement by line managers in systems development. Anyone who has had a long spell in systems design since leaving school/college/university without any line experience can be highly dangerous.

(2) Companies have lost or won the business battle on how they have tackled their product markets, carried on production or handled their employees. All these are important. It is just possible in the 1990s that information may be more important than them all. The organization could be geared around the use of information—its generation, its storage, its manipulation, its retrieval, its communication and, above all, its use.
(3) Information, then, should strongly influence how a company is organized, how the variety of potentially contending managers and specialists can unite to form effective working groups, how objectives can be set and achieved, how the company relates to its customers.
(4) A hard, demanding look at the costs of IT and the benefits actually being achieved needs to be made first, as if it were a production machine being put into the factory.
(5) Without good systems design, IT will prove abortive. The two most important systems a manufacturing company needs are MRP II and Management Accounting. All companies need the internal discipline that these systems can give.
(6) Systems should be used to help weld the organization together. They should be all-pervading, being used by managers, supervisors and, if necessary, shop stewards alike. All should understand them and be able to use them. Training should teach the systems and also ensure total commitment to them.
(7) Top-down planning should be the basis for setting budgets and budgetary control. The Management Accounting system should be oriented to allow this and a subsequent debate. Bottom-up planning will nearly always give a less than optimum plan. It will fail to ensure that managers are stretched.

A suitable set of strategies

(1) Our IT strategy will be to exploit the technical hardware which is currently available, to help achieve the overall objectives of the organization. As much hardware as possible will be placed in the hands of line managers. The Systems/EDP department should never be 'the black art' practitioners. All IT projects will continue to be judged on their cost/benefit irrespective of the nature of the project.
(2) We will establish systems networks, integrating wherever possible separately developed systems, but never allowing new systems to be developed which cannot form part of a network.
(3) We will continue to develop the company database to allow its access by all nominated people.
(4) We will continue to push as much systems development activity as possible into the hands of line managers. We will reduce the systems department wherever this appears to be justified.
(5) Emphasis has changed from tough evaluation of hardware into well-considered judgements on software. We can only see this trend continuing. Cost of software and the actual software developed will be very closely monitored.
(6) With reluctance, we have decided to abandon many stand-alone systems in favour of an integrated MRP II activity. Throwing away years of careful systems development is very painful, but is essential.

(7) As far as possible, we will concentrate on open systems which can be run on a variety of hardware. We should never design systems which lock us into one supplier. We will use a UNIX operating system to achieve this end.
(8) We will continue to use system resources to ensure that MRP II is effective. We must be a Class A user. We carry out this activity at the expense (if necessary) of other systems development, with the exception of Management Accounting.
(9) We are in business to sell production capacity to the highest contribution bidder. Our MRP II and Management Accounting systems should be oriented towards achieving this objective.

7

Motivation and Reward Systems

7.1 THE DILEMMA IN PAY AND MOTIVATION

What employees in a manufacturing company can be paid, or indeed should be paid, has been a bone of contention between owners, managers and workers since the beginning of the industrial revolution. Achieving a satisfactory relationship between effort, productivity, competition, external economic forces, profit and pay, has been, to say the least, difficult. Bitter strikes and the death of companies are only two of the possible adverse outcomes which have been all too much in evidence.

The position in the UK can be a real and debilitating one. It is far too easy to say that pay should largely reflect what profit/contribution/added value a company is able to achieve. The real world often produces a different result.

Most workers in the UK are now employed by organizations which do not have to face international competition. Charge capping, cash limits or even privatization may threaten people in the National Heath Service or local government, but it is extremely unlikely that their pay and productivity will in any way need to relate to what workers in Taiwan or South Korea achieved or are paid. The service provided by the NHS is not normally undercut by cheap imports from Third World countries.

If local government workers in Matlock are paid an extra 8 or 10% by Derbyshire County Council, why should operatives in a nearby motor components factory receive only 3% because the company needs to remain internationally competitive?

There is more than superficial wrong in a situation where, in the same week, a production manager receives a pay rise demand from his operatives which matches that just given to the local leisure centre workers and an increase in the site rates bill which will fund the leisure centre people's pay rise.

Why should the same motor components manufacturer pay more for its electricity than its competitors in Germany, because the electricity company has a quasi-monopoly over supply and within limits can charge what it likes?

With, say a 5% inflation rate in the UK and one of 3% in Germany, the only way a UK manufacturing exporter will survive is to ensure that:

(1) The exchange rate adjusts to take account of the difference in inflation rates. With ERM entry this is no longer an option. It is essential therefore that national/external economic pressures will only equal other nations' rates of inflation.
(2) Productivity improves so that price rises based on German inflation cause no problem.
(3) Pay is kept down to levels which match those in Germany.

As well as the national economic problems which have had a major influence on pay and profit, there are other, more internal ones:

(a) Organizational ineptitude
It is possible that organizational incompetence could compound the pay dilemma. For example, a company's profit plan normally runs from January to December (i.e, the Financial Year). The plan is usually firm in October of the year preceding the profit plan period. An important element in the plan is the increase in pay which is anticipated (if any). Wage bargaining may begin in May and June when the profit plan has already run for six months. Economic conditions may have changed considerably by then. Whoever was responsible for forecasting the possible wage increase could have been very wrong.

For many years in Ferodo's case, it was not thought necessary for the Personnel Manager carrying out pay negotiations, to know of the pay increase figure put into the plan. Apparently the Personnel Manager would do his best to achieve the smallest rise he could get the unions to accept. No one debated the harsh results of paying too much or not achieving adequate productivity. Mainly this was because it was thought that such factors were not part of the pay negotiating activity.

Initially the line managers who would have to suffer the cost changes brought about by 'excessive pay rises' were not included in the negotiating team. If the main function of the Personnel department was compromise and its main activity discussion, it is little wonder that inadequate pay discussions took place.

(b) Union greed?
Senior managers in manufacturing are well aware of many of the unions' demands over pay:

(1) Pay should rise to cover inflation. 'The lads' need to pay their mortgages. Some companies have made settlements with their unions on the basis of 'inflation + x%' (e.g, Ford Motor Co.) at the end of the 1980s. Without cost containment or productivity improvement to offset this kind of deal, disaster can strike.
(2) National trade union organizers demand that their local shop stewards gain a standard pay rise which the trade union head office has set as a norm. (The MSF Union appeared to be playing this game in the late eighties.) The norm has to be paid irrespective of the economic state of the company.

(3) The workforce themselves see local service industries or local government people gaining 'X'% rise and demand the same.
(4) Obvious signs of profit and productivity improvement resulting from investments in high technology and paid for by borrowing from the bank are used as an excuse for a pay rise.
(5) Relative deprivation drives some part of the workforce to demand parity with another section or work organization, even though all current pay rates really reflect job worth and what the company can afford to pay.

Good manufacturing management should accept none of these reasons for giving pay increases.

So, are British unions so greedy and stupid that arriving at reasonable, productivity-related pay increases is impossible?

It is not too difficult to sympathize with a workforce which is told nothing about how well or badly the company is coping with competition; how little or much profit is being made; which investment plans are in process or, perhaps with more direct relevance, how the pay of the local workforce compares with that of other companies in the area, or the local Town Hall.

If along with the absence of such information there is a glaringly obvious gulf between the employment condition of workers and management, in pay and pensions, canteen, clocking or not clocking, even in where cars are parked—then there is bound to be friction and quite likely aggression over pay demands. Pay tends to be the one battle that the unions can fight, which apparently redresses obvious discrepancies between management and the shop floor. Why not push it to the limit?

So, in Ferodo's case, we spent three or four years giving out information, fully supported by finance courses based on the last ten years of company results. Only then did the senior shop steward of the Transport and General Workers' Union write to his members in his annual report to his branch, as follows:

> Surely no one would wish for a return to the situation which existed not too long ago, when we were enjoying wage increases paid with money the company had not got. This was responsible for bringing all of us to a situation which led at the time to redundancies, wage cuts and the critical times when a receiver was appointed and the bank was in fact preparing to 'pull the plug' on any further financial support. It is fair to say that many people at the time just did not believe the situation was so bad. But thankfully common sense prevailed and it would appear that the sacrifices made at the time have to large degree led to greater stability and security reflected in increased work load and more job opportunities.

Of course there are unions and their shop stewards who will aggressively defend the status quo, obstruct the introduction of new technology, and protect obsolete working practices. The chances are that senior management is equally conservative; but without dialogue and understanding of the company position and performance, pay negotiations can be nothing but a battle to get more than the company can afford.

7.2 IS MONEY THE ONLY MOTIVATOR?

It is probably true that most production managers still believe that money is the only motivator, or at least the main one. In an age when materialism appears to be increasing, it is an easy assumption to make. Yet there is quite a case to be made for the opposite viewpoint.

(a) Senior managers are quite happy to accept bonuses and share option schemes based on their performances. However, they would resent a belief that they are a little like Pavlov's dogs, responding in a totally mechanistic way to a stimulus. Yet this is how the traditional payment-by-results schemes, which operated on many UK shop floors, treated operatives. Why should there be a difference between the two sets of people?

(b) It seems possible that shop-floor workers can be motivated by things other than money. Most managers will have heard, from their very early training, of the 'Hawthorne Experiment'. Taking notice of the workforce seems a good idea.

(c) Norms or standards of output, especially among teams, are common in UK manufacturing. No one works so hard that he/she 'spoils the job' for the rest. There are limits set, by a common understanding, on output, irrespective of whether more could be earned by producing more.

(d) When money is used to attempt to motivate, it is occasionally counter-productive. In Ferodo, first-line supervision was at one time not paid for overtime. Overtime consequently was limited. At some time, an agreement was made to pay for extra hours' work. As a result, overtime increased enormously. Payment for overtime had reduced the status of this group of people They responded accordingly.

(e) In any factory there are quite wide-ranging discrepancies between effort and reward being achieved. Payment schemes can differ of course, but the biggest difference probably lies in the age and general composition of the local teams, their conditions of work and the effectiveness of local management.

(f) How often does it seem that people on the shop floor are more concerned with stability of earnings than in maximizing them?

(g) It is interesting that within the same locality, companies can pay widely different rates yet not achieve significantly different results. Companies which have strikes and labour unrest are not always those that pay the least. Often the converse is true. There must be something else which acts as a motivating or de-motivating factor.

So although these may not be totally conclusive, it does seem that using money as the sole motivator is wrong.

7.3 NON-MONETARY MOTIVATION

Even many hardened production managers once considered some of the approaches to motivation proposed by industrial psychologists. Herzberg, Maslow, McGregor were well known. Herzberg[†] in particular became something of a guru, putting forward views on job enrichment.

There were various factors, we were told, which *we* thought would motivate, but which never would, such as:

(a) More pay.
(b) More holidays and a shorter working week.
(c) More fringe benefits such as pension schemes, sick pay, time off for family problems, etc.
(d) Communications.
(e) Sensitivity training (saying 'please' and 'thank you' when you talk to workers).

For Herzberg, the only salvation lay in job enrichment. Jobs, he said, would be restructured so that individuals are given greater responsibility for the work they perform and especially in recognition of achievement. People should be given enlarged jobs, where new and perhaps more difficult tasks are performed and where self-control is paramount. Acquiring new skills should be high on the agenda.

Various organizations round the world have attempted to introduce job enrichment. Volvo in Sweden was a pioneer of the process. Yet the degree of success appears, from all the literature available, to be sparse. Why?

Restructuring jobs, purely to provide job enrichment, is obviously a process which few will embark on without a degree of cynicism. It is hard to believe that a semi-literate operative will respond to job enrichment. All his attitudes are wrong for the process. Something needs to be done prior to job enrichment processes being introduced.

The ending of dull, monotonous, soul destroying jobs is itself a useful goal, but not if it does not raise productivity. Many factories use technologies or have processes where technology alone cannot provide the requisite outcome. Some jobs defy job enrichment.

Job enrichment has rarely been asked for by trade unions. It seems to have a pretty low priority for them, coming well after pay and job security.

'Work itself', one of Herzberg's motivational factors, has always been tied up with salary, job security and personal relationships. It is difficult to separate out 'hygiene' from motivational factors.

Why people work is a useful debating point. Obviously because they have to due to economic necessity; but work also provides two other valuable factors in people's lives. Firstly, it gives social contact. Secondly, it provides a degree of self-worth. The job itself enables people to feel a degree of pride and importance—or it should.

So does it appear, out of all the senior common room debate among social and industrial psychologists, that there is one factor worth pursuing or at least keeping in mind when considering motivation? Self-worth could provide part of the answer.

[†] See, for example: F. Herzberg 'One more time—how do you motivate employees? *Harvard Business Review* Jan– Feb. 1968.

Self-worth

Even directors, blessed with a good bonus scheme and a pocketful of share options, probably see self-worth as a true motivating factor. Their jobs give them satisfaction. They feel the better for holding them. Increasing their self-worth is a stimulus for action. There is little low self-esteem, unless the Chairman starts to put his weight about.

Unfortunately UK society, as a whole, seems poor at generating self-worth among a large proportion of the nation. Fathers are unemployed or carry out an unskilled job. The educational system seems to turn out many under-achievers. Day-by-day pressure makes them defiant of authority, but also incapable of running their lives in a disciplined way.

High self-esteem appears, according to reports, to generate some pride in achievement causing people to do a reasonable job when asked and to dislike those who behave in an anti-team or anti-organization way. The number of occasions when shop-floor people went out of their way to congratulate management on taking action against known malcontents increased, in our case, the longer we attempted to introduce 'self-worth'. Eventually, there was a threat of a strike if well-known rule-breakers were not punished.

People want to belong to a successful team, to have some self-respect and pride in the job they do. They want to measure their success in some way other than achieving as much pay as possible for as little effort as possible. Why are factories run in a way which does not provide this motivation?

Generating self-worth can be difficult and far from painless, but it can be comparatively cheap, compared with putting in large traunches of new investment. The following measures are recommended:

(1) Treat people as if they matter. Give them the best possible working conditions. Give them the best canteen. In Ferodo we staged a constant battle over housekeeping. There was a constant nag, 'Would you treat your own home like this?' That some might, made no difference to the push to make the shop floor clean and pollution-free. Time spent on housekeeping was never wasted.
(2) Spend time and money on training, especially in statistical process control and quality generally. Everyone should go through a training course. Everyone should be given the idea that they can be better than they are.
(3) Everyone should belong to a team which they can relate to and identify with. The team should be set important goals and be told how well it is achieving them—daily, weekly, monthly. Focused factories, business and product lines, all help in this. A conglomerate factory is not a good idea.
(4) Let team members have some say in its structure, how its performance can be improved and generally how things are done.
(5) Teams and groups should be small enough to gain close identification. The Romans had some good ideas about the structure of their army and standards to which the soldiers could relate.
(6) Communicate. Let operatives take time off to be told about their group's performance and that of the company as a whole.

(7) Give praise when due. Reward good overall performance as well as individual performance. Build payment schemes around this.
(8) Abolish as many privileges as possible which make many people believe that there is a class divide and they are on the wrong side of it. Eliminate anything that can possibly promote 'us and them' attitudes, such as:

– separate canteens
– clocking in
– varying hours of work
– company cars.

One public performance tends to distinguish many British-run factories from those with Japanese parentage: It is the speed at which British workers leave their factory at shift change time. When at least some of the team members stay behind to discuss what went right and what went wrong during the shift, production managers will know that their 'improvement in self-worth' campaign is starting to have an effect.

7.4 INCENTIVES AND MOTIVATION—AN APPROACH

To start our own debate on the future pay strategy we might deploy, we wrote a position paper, as follows (sections 7.4.1–7.4.4):

7.4.1 Background

(a) No matter how hard we try, we seem unable to make substantial improvements in factory pay compared with £1 of revenue earned.
(b) The use of Bedaux 60/80 incentives appears to do little to improve quality, reduce rejects, get products out on time, maximize machine/equipment utilization, or keep the place tidy and healthy. They do, however, sustain a degree of output discipline, without which performance could be significantly lower than it is now. Eliminating such schemes without a satisfactory alternative would be a disaster.
(c) The incentive schemes generate a culture, or attitudes which cannot be appropriate for a manufacturing unit in the UK in the 1990s. One worrying factor is the complexity of the incentives and the wage payment systems, which allow wages-drift.
(d) First-line management has used the 60/80 incentive schemes to ensure that output of some degree is achieved. They have sheltered behind such schemes and rarely, if ever, actually managed their section or parts of the factory for which they are responsible as well as they could have done.
(e) Most of the older members of the workforce have worked all their lives with the current incentives. They have always been reasonably well paid and have been able to manipulate their pay to suit their needs. They see little need to change, especially if it takes away their ability for self-created pay.
(f) The younger members of the workforce quickly pick up the failings of the old, because they have nothing else to which they can become attached.

(g) Our own pace of change is slow. Payment by programme schemes should be introduced much faster than we are currently achieving.

7.4.2 Requirements

(a) Incentives motivation
These should be:

– simple and straightforward
– easily controlled
– related to all aspects of the business, not just to output in pieces or reward purely for effort.

(b) Work organization
• Autonomy within well-defined and controlled parameters
• People able and willing to do many jobs and move around to do so
• Section managers to be 'facilitators' as well as managers
• Effectiveness of sections or work organizations to be measured in:

– output in pieces
– rejects, material utilization, quality
– output on time
– unit costs measured against standard
– machine utilization
– contribution earned
– added value earned
– pay to added value.

(c) Possibilities
Where current incentives cannot be eliminated, they should be stabilized and simplified. Control should be improved considerably from current levels.

New incentives should be based on reducing unit costs, not reducing pay. The cost of rectification of poor quality, or extra business cost incurred though operative failings, should be set against pay.

Where current 60/80 incentives need to be retained, an overlay payment should be negotiated for well conceived work organization.

The overlay should be based on contribution, added value or some other business measurement.

Section managers and other production managers should have their own pay related as closely as possible to the performance of the working groups under their control.

The essence of the new schemes should be to improve business performance in the widest connotation possible. We need 'autonomous work groups' and associated motivation schemes.

Payment by programme still seems the simplest way of controlling what is done on the shop floor.

7.4.3 Needs
We will never succeed in changing factory incentives unless:

(a) first-line supervision is fully trained and motivated to play a significant role in improving factory performance. We have started, but still need to do more.
(b) the workforce and especially shop stewards, see that we are changing our culture. We have started, but again, much more needs to be done.
(c) group working performance is measured accurately. We are counting rejects and output better, but it is still not good enough. Cost control is spreading too slowly.
(d) work organization is satisfactory. We have gone some way towards splitting up the factory and measuring it. We need to firm up on this new organization and ensure that the workforce is as flexible and autonomous as possible.
(e) We use non-money rewards as well as money rewards in achieving good total performance.
(f) the workforce needs to be educated to understand business needs and to appreciate that future prosperity and jobs depend on their relationship with business performance.
(g) MRP II works effectively.

7.4.4 End result
The end result should be:

(a) a significant reduction in unit costs. The payroll/added value ratio should improve by at least 3% for each of the next five years.
(b) considerable improvement in the quality of products made.
(c) considerable improvements in flexibility and general appreciation of the company's business performance and what is needed to improve it.

7.5 PAY SCHEMES—ALTERNATIVES AND PROBLEMS

In reviewing our own pay procedures and what might be a better substitute, we reviewed both 60/80 payment-by-results and as many alternatives as possible, ranging from measured day work to Rucker/Scanlon added value systems. None seemed ideal. In the course of our research we came to the following conclusions.

(1) How people are paid often reflects the ability or desire of management to manage. There are factories which are totally dominated and controlled by payment-by-result schemes. In a 'leave well alone' situation, first-line supervision can become progressively less effective, reduced to being conciliators and supporters of the workforce.
(2) The British have seemed particularly keen to deploy PBR. The Unions normally treat it as a football in the great game of beating the management. Shop stewards and operatives, to a degree, can delude the inexperienced work study man and exasperate the experienced, so much so that an unsatisfactory outcome for management is nearly always the result.

Few factories which use Bedaux 60/80 schemes seem immune to exploited values, wages-drift and a constant knocking at established values in attempts to get them changed to the operative's benefit.

(3) PBR is normally associated with individuals or small teams. A broad outlook is not promoted. Once standards have been agreed, operatives have a vested interest in keeping the job as it is. Change would normally be detrimental to the operative, so it is resisted.

(4) Controlling PBR wage payments effectively in a factory where, say, a hundred or more schemes are in operation is difficult if not impossible. Inevitably, work booking becomes suspect and operatives adept at 'working the system'.

(5) PBR often restricts output rather than increases it, as restrictive practices are built into the system. The concentration on effort and pieces produced often does little for other key cost factors, such as material utilization. Unions have become adept at demanding—and receiving—pay, even for poor quality work, because of the difficulty of pinpointing who or what is responsible.

(6) Yet despite these drawbacks, PBR on F.W.Taylor's principles still seems widely applied in British industry. The reasons?

- Inertia
- Work study practitioners know no better
- Production management has not thought through the benefits of changing from PBR
- Junior management knows no other way (or indeed has not been trained to accept another way) of running the factory
- The culture needed to operate another system of reward successfully is missing
- The unions like PBR because it provides endless opportunities for debate and profitable change.

Other schemes have been tried:

Added value
Profit sharing
Scanlon–Rucker
Multi-factor
Measured day work
Productivity/inflation linked—using appropriate indices.
Long-term pay deals.

Added value, sales value of production and cost improvement have all been used as the basis for incentives. (Added value is defined as 'the value added to materials and other purchased items which provides, as a result of productive activities of the firm, the sum out of which salaries and administrative overhead expenses are paid, leaving any surplus as profit'[†]).

[†] Engineering Employees Federation.

We were keen to use added value, but found that the understanding of it was very limited and would have necessitated considerable training. Other factors were also negative:

- Capital productivity—At a time when considerable capital expenditure was being made, it seemed unwise to base an incentive on:

$$\frac{\text{Added value}}{\text{wages paid}}$$

- Effort—Added value does not measure 'effort' as such, only the reward of effort.
- Pricing—Added value is very susceptible to price rises or indeed to the failure to achieve adequate price rises.
- Culture—We still had some way to go to change the culture of the organization, from which added value incentives could derive.

The Rucker and Scanlon plans are not new and have been used in the USA for many years. It is significant how many such schemes have been tried in the UK and then failed, reversion to Taylor-type PBR being the most common outcome. This says much about the training of management and the attitudes of the workforce and, above all, the culture of the organization.

Our own options were reasonably clear,:

(1) Continue with Taylor-type incentives with all their drawbacks and deficiencies. At least they gave a degree of discipline and output for money paid. Union pushing for better rates and conditions could be containable within limits. However, if we wanted to increase the cake, get better co-operation on a range of things other than output, indeed, get more like the Japanese, then staying as we were was not an option.
(2) High day rate. Where machines dominated operatives, as on the robotic cells, high day rates seemed a possibility. There would be no incentives, as such, just reasonable pay. We had chosen the best operatives for these jobs, but they became increasingly dissatisfied as they saw wage-drift in other jobs erode their special position. High day rate and Taylor-type incentives do not mix in one factory.
(3) Measured day work. This meant abandoning piece work and introducing a bargain or contract with the local operatives. For an agreed amount of output a guaranteed pay level would be established. A reasonably tough, even aggressive local management could be needed, which in itself might be counterproductive. Entrenched culture in favour of PBR would certainly not allow an easy transition.
(4) Current system plus overlay for product line or site performance. This seemed an attractive proposal if the overlay element gradually became the dominant part of the payment system. The Current 60/80 scheme could be allowed to wither and die. However, it did not solve the problem of how to achieve flexibility, team working, etc.
(5) Group incentives. These had been tried. Some schemes had worked, others had not. They worked better when local first-line supervision was effective and reasonably well trained and motivated.

All the alternatives we considered produced at least one common thread. It has been possible over the years for direct operatives to manipulate their pay so that they achieved the level they thought appropriate. If values were slack, it meant that extra time could be spent in smoking cabins, gossiping or filling in football coupons. They were being rewarded for their immediate effort. Whatever activity went on in the world outside, in low prices, harsh competition, the need for ever-increasing quality standards, passed them by. This had to end. Whatever change was introduced, it would not come without a lot of pain, considerable acrimony and debate, plus compromise on all sides.

At least we were now sure of what kind of objectives we were trying to achieve in changing the payment system:

(1) To gain an ability to raise productivity, lower unit costs without the addition of new equipment and to do this on a year-by-year basis. To get the active cooperation of the workforce in getting better results. To ensure that peer group pressure was positive on those who did not want to perform.
(2) To end the isolation of the shop floor from the outside world, especially in the demands on cost reduction, better service and improved quality.
(3) To reward individual skills on the same basis right across the company.
(4) To end, if at all possible, relative deprivation.
(5) To minimize administrative costs.
(6) To improve industrial relations and the culture of the factory.
(7) To end wages-drift
(8) To provide a facility for everyone to work within a team
(9) To improve maintenance of equipment and the speed at which stopped machines were repaired.

7.6 MANAGEMENT INCENTIVES/PAY SCHEMES

While PBR has, apparently, been appropriate for shop-floor operatives, management has often achieved pay awards comfortably ahead of inflation. For those who survived the spate of redundancies in the early eighties, the later years of the decade were fairly golden times with a good deal of catching up going on. Increased profits have been followed by higher salaries, even if the two have not been directly related.

Organizations like Ferodo, that have tried to relate managers pay and annual pay increases to some form of job evaluation and performance appraisal, have only partially succeeded in controlling total pay rises. These schemes depend upon a well-organized points rating system and, subsequently, sound merit rating.

Anyone familiar with merit rating knows that most managers will mark their subordinates as satisfactory or above, even though it is quite apparent that one or two of the people being appraised deserve to be marked quite badly. Rarely if ever does the result give statistical regularity. There is always an excuse that 'X' and 'Y' deserve more because they might leave, or 'Z' is now trying harder and it would be wrong to demotivate him.

How to write a job description so that it get the maximum number of points is an art form which managers, in self-defence, rapidly teach themselves.

7.7 RELATIVE DEPRIVATION AND PAYMENT SCHEMES

This is the second time relative deprivation has been put forward as a potent problem in establishing both satisfactory work organizations and pay that goes with them. British industry may not have a pay problem as such. It has a problem over rewards and status.

Most shop-floor workers are better off than they were thirty years ago. The possibility of owning a car and a house, having reasonable holidays and healthy well-fed children has improved immensely. Yet there are still strikes over pay. There is still resentment towards office or management personnel who might be less well paid than operatives are, but who have much better working conditions, pension arrangements and staff status, etc.

Conversely, staff and management personnel might consider shop-floor pay as too high and too easily earned, with a basic lack of responsibility in how work is carried out.

This is relative deprivation at work. All people may be comparatively well paid but when they compare their job and what it entails they often believe that others are treated better in many ways.

Why should a newly joined sixteen-year-old trainee word-processing operator have paid sick-leave, a better pension scheme and perhaps a better canteen to eat in, than someone who has worked on the shop floor for thirty years?

Pay cannot be divorced from a whole range of items which could cause relative deprivation, such as the following:

- Hours of work/shift patterns
- Holidays and when they can be taken
- Ability to take time off or work other hours, when, say a domestic crisis occurs
- Physical conditions—noise, heat, dust, exertion
- Clocking on and the disciplinary action taken for lateness
- Weekly/monthly pay
- Bonus payments and ability to achieve extra earnings
- Prospects of promotion/upward mobility
- Canteen and other welfare facilities—rest rooms, etc.
- Period of notice prior to dismissal
- Sick pay, and how long periods of sickness are paid for by the company
- Degree of supervision and individual responsibility for work carried out
- Amounts of pay given.

To follow Herzberg,[†] these factors break down into hygiene and non-hygiene factors. Not all of them are detrimental to shop-floor workers.

[†] Herzberg *Work and the nature of man*. World Publishing Co. 1966.

It is common for younger people with families and a mortgage to put up with poor working conditions in exchange for the opportunity to earn relatively high pay, with bonus and overtime. As a worker gets older, he tends to have fewer financial commitments and he might give social activities and leisure time priority over high earnings. (Selling new or revised payment systems is often influenced by age-differences.)

Age differentials matter considerably in how workers regard relative deprivation, whether they will work overtime or what physical conditions they will accept. The Japanese are probably right to recruit only young people in their twenties, if they want to achieve high output, flexibility and a corporate ability to change.

Many surveys have been done which suggest that professional people have the highest job satisfaction, followed by managers, clerical people, skilled tradesmen and finally semi-skilled or unskilled shop floor workers.

When Ferodo introduced staff status for the engineering staff, it seemed that we were introducing a considerable degree of job satisfaction. Some problems we had were as follows:

- How to institute a degree of self-discipline into a body of people who for decades had been dragooned by clocking on and off.
- How could abuse of the more relaxed discipline be prevented?
- Once pay is given for all sickness, or even general absence, what regulating procedures are needed? Some obviously will be necessary.

Junior management and first-line supervision are likely to resent—in part—the levelling up that single status brings. Two possibilities could follow. Firstly, there is a keen desire to see that no abuse occurs. Secondly, the managers demand further pay and status for themselves to ensure that they do not suffer by comparison.

It is difficult, if not impossible, to make a factory as pleasant a place to work in as an office, but it pays to try. Regretfully, it is often a fact that the British appear singularly unwilling to keep either their streets or their places of work clean and tidy. Ferodo had always put cleanliness and health and safety at the top of the list of key objectives.

The answer to relative deprivation, therefore, is single status and the best possible working conditions for everyone—not a cheap or an easy solution.

7.8 OBJECTIONS TO CHANGE AND POSSIBLE SOLUTIONS

To attempt to change a pay structure without the most detailed and lengthy discussion with shop stewards and the general workforce will inevitably cause immense trouble. It is one of the most emotive of all subjects.

Talking to the shop stewards and gaining their acceptance can be a long way (in time and place) from getting general agreement right across the shop floor. Single group schemes can be sold (eventually) by one-to-one contact, with all the people who will take part in the scheme. A company-wide change is a very different and immensely more difficult proposition.

Initially, any bonus or incentive based on site, or even product line, profitability received (in our case) short shrift. Profit, the operatives said, was not produced by them, but concocted by management. Why should operatives depend upon their machinations for their pay?

We met this fairly unsatisfactory opening debate by trying the following:

(1) All shop stewards, associated trade union national officers and our own managers were brought together to hear what kind of a company we needed to be to survive in the 1990s. (This is the report given in Chapter 1.)
(2) We called in consultants, who wrote a report on current failings and proposed changes. The report told management very little not already known and followed guidelines we had written (section 7.2). But there in all their gory detail were listed the 'red circle' jobs that had to be brought into line and which were receiving pay well in excess of what any fair and rational payment scheme would produce. There, too, were the blunt conclusions about what failings were occurring and what needed to be done to put them right. Again the proposals were broadly what we would have written anyway. Copies of the report were issued to the shop stewards and national officers. While this gave the stewards an opportunity to rebut some of the more contentious statements, it gave a clear indication of what reasonably well-qualified outsiders thought of the company's payment system.
(3) Managers were put into a hotel for the weekend to discuss the report and propose their own solutions to current pay problems Debate turned on these aspects:

 - Gaining union agreement for whatever was eventually proposed.
 - How much it would cost to eliminate 'red circle' jobs.
 - What was still necessary to change the culture of the shop floor completely.
 - What needed to be done to make team incentives effective.

Occasionally, even with good and developing management—shop steward relationships, a rogue bull is needed. We decided to cost out the establishment of a green field site in the North East of England, producing most of the high contribution products we made. The report was electrifying for management and shop stewards alike. A summary of the findings actually found its way into the *Financial Times*. What manager at some time has not wished that his organization could start all over again, eliminating all past mistakes, with full authority to negotiate brand-new pay systems with a newly recruited workforce. We certainly had. We envied the Japanese starting on a green-field site, with one or even no union agreement, with an untainted workforce keen to do its best.

We challenged our own unions to accept, out of a sense of social responsibility, change, which would result in at least a 'brown-field site', if not green.

The unions thought this rogue bull was unfair and unnecessary pressure on improving relationships, but nevertheless it helped to create a climate where change was possible.

Everyone on site was issued with an information booklet which recorded the challenges facing the company and what needed to be done to overcome future problems. The simple illustrated pages did not appeal to everyone. Some illustrated items were:

- What management was doing wrong which needed to be put right e.g.:
  - first-line supervision - training and motivation
  - quality/SPC
  - cost control
  - communications
  - flexibility

- Shop stewards and especially when to involve in cross-union working
- Customer service—quality, deliveries, costs, changing market structure, operative impact on customer service
- What was a team?
  - its composition
  - break points

- How to achieve change:
  - communication
  - organization.

7.9 MANAGEMENT INVOLVEMENT IN CHANGING PAY SYSTEMS

Junior management was highly suspicious of any changes in payment to an already well-paid workforce, without substantial improvements in performance. So before any 'hearts and minds' campaign was started with the unions, management needed to be won over to the radical changes which were needed. This included support for the 'managers right to manage', even though this at the time seemed retrograde. However, it did appear to provide a platform from which senior and junior managers alike could view change and payment with equanimity and reasonable morale. Among the 'right to manage' and morale-raising activities we highlighted the following:

(1) All agreements where 'flexibility' had been bought as part of a past agreement were to be enforced rigorously.
(2) Discipline on the shop floor was to be tightened as much as possible especially for work badly carried out.
(3) Where possible we began to relate first-line supervision's pay to the achievement of their group.
(4) The Union *bona fides* were tested continually, especially in getting shop stewards and local management together to solve common problems.
(5) Wage payment control was toughened up.
(6) Overtime which had become institutionalized in parts of the plant was cut back and controlled carefully.
(7) Communications were improved and extended.

The process of change and adaptation was helped by putting in more 'payment-by-programme' systems. These were based on the work programmes which were issued for scheduling/planning purposes. Each programme was allocated a standard time for its completion. This time was the aggregate total of work study values/times already accepted by the Unions. Once the programme was completed, the standard payment was made to the personnel who actually operated upon it.

Like all group payment systems, the allocation of the reward among the group caused some anguish, particularly between shifts. Some shifts worked hard, others did not. A payment which rewarded all shifts equally was patently unfair.

Paying individual shifts was a retrograde step which the differential in shift output forced into operation. However, opportunities for inaccuracies or malpractice in work booking largely ceased. Operatives began to see products as an output from their effort, not 'earned work study credits'.

Payments-by-programme also helped in the slow change in factory culture. Debate about the programme, its work content and its achievability (within the time allowed) was held at the start of the first shift each Monday. The traditional role of the shop steward began to change.

7.10 THE WAY FORWARD

7.10.1 Introduction
The *Financial Times* of 23 December 1988 reported on the decision we had made:

Ferodo seeks greater flexibility to save jobs
Ferodo .....has told its workers at its Derbyshire plant that they must accept widespread changes in working practices.

A number of manufacturing companies have made similar initiatives to try to introduce new practices—often breaking down traditional demarcation within existing plants. They have become known as brownfield sites.

The company said that there would be a review of working methods in all departments. It believed there could be gains by introducing multi-skilling in some areas and eliminating demarcation between grades.

Present job definitions, grading, pay structures and production agreements would all have to be changed.

It is essential that we optimise the use of expensive capital equipment, learn new skills and become more flexible. It is intended to broaden jobs, provide opportunities for better career progression and improve earnings.

Ferodo did not intend to divert capital investment away from its Derbyshire plant, although it had considered alternative sites in a review of operations. It preferred to raise productivity with Union agreement.

7.10.2 Basis for change
We recognized that change in systems had to be brought about on the following lines.

(a) We needed to build up rewards for performance of teams, product lines and sites.
(b) We needed union cooperation in actually designing the systems, not just in agreeing to them.
(c) Rewards for age, seniority, service were to be strictly limited, if considered at all.

(d) Competence needed to be extended over a wide range of activities especially in maintenance, tool changing and machine alteration; i.e., multi-skilling was essential.
(e) Elimination of individual job descriptions, to be replaced by factory-wide job evaluation.
(f) Pay and pay increases had to relate to the company's ability to pay.
(g) We needed common terms and conditions.
(h) All payment systems needed to be integrated.

To achieve these aims we obviously needed to:

(1) Rationalize jobs, write suitable job descriptions and train a panel of people to carry out job evaluation based on agreed bench-mark jobs. Then ensure that all jobs were fitted into the structure.
(2) If team incentives were to be used, then performance measurements for the teams, based on input–output data were needed. Standard hours seemed an appropriate measurement with a reduction in standard cost and standard hours per product. Clocked hours compared with standard hours achieved also seemed a valid basis for incentives.
(3) Ensure that every step we took would be self-financing, even if we had to use consultants. Largely, our own internal resources were required.
(4) Retrain—especially members of the CSEU and first-line supervision.
(5) Ensure that nothing happened that would reduce:
   – output
   – quality standards
   – health and safety standards.
(6) Improve all the ancillary activities which impinged on effective team working:-
   – Planning
   – costing and cost control
   – supervision
   – training
   – use of relief operatives.
(7) Re-validate work standards, to arrive at an accurate standard hour per product, but at the same time simplify the process of establishing standards.
(8) Review all operations where operatives were dominated by machines and ensure that the incentives include maximization of machine utilization.

7.11 SOLUTIONS

Four solutions to our problems of pay and motivation were agreed.

(1) *Team incentives.* Set up autonomous work groups, as incentive and work organization go together. Make sure that the mixture of CSEU, T&GWU and MSF people was established in such a way that inter-union squabbles did not occur.

The teams should therefore comprise direct, indirect and maintenance personnel, plus supervision and clerical support and, where absolutely necessary, technical personnel.

The teams' incentives should be based on a reduction in unit cost of output compared with labour and material cost of input. There could be a transitional period when the group incentive would form an overlay to individual bonuses on a basis already in operation. There would be a definite period during which the individual bonus could wither and die with the possibility of having site, business or perhaps product line bonus incentives.

(2) *Integrated grade/payment system.* This was to be established for all people below management grade, irrespective of union, occupation or current status. Both 'white' and 'blue' collar workers were to be involved. Emphasis would be on:

- Extra pay for multi-skills and flexibility.
- Few grades—perhaps four at the most, with minimum and maximum pay scales.
- Everyone having a pay rise based on company performance.
- Establishing company-wide job evaluation, on which 'integrated pay' would be anchored.
- Solving relative deprivation problems, especially between technically qualified juniors (say, in R&D and Quality Assurance) and shop-floor operatives.
- Ensuring new technology was properly handled with appropriately paid individuals.

(3) *Harmonization of terms and conditions—as a matter of principle.* We had already made a start with the CSEU and this approach needed to be extended to the T&GWU people. A major problem could be in pension funding.

(4) *Second-tier payment.* The best basis for a product line payment appeared to be:

$$\frac{\text{Net revenue earned}}{\text{Interest paid on average WIP and FGS} + \text{Factory cost,}}$$
with raw material valued at standard

It probably seems unreasonable to include finished goods stock (FGS), but when 'just-in-time' is being pressed then all inventory is important. Finished goods stocks do have a relationship to factory cost in that producing larger stocks has probably meant lower factory cost. This solution has the merit of being simple, relatively easy to understand, fairly easily calculated, yet strongly related to company success.

7.12 FUTURE CONSIDERATIONS

Some companies, especially those with Japanese or American parents (e.g. Texas Instruments, Komatsu, Johnsons Wax) have developed strong links between performance and pay. Skills acquisition is a key factor and pay is related to it. British-based companies seem less keen on this approach; probably their unions do not want to give up a rate for a job. However, the way is clear.

In Komatsu for example, there is a six monthly appraisal.

Companies who have installed performance appraisal and pay might use the following factors:

Attendance
Timekeeping
Productivity
Quality
Dependability
Job Knowledge
Flexibility
Good housekeeping
Performance against objectives
Teamwork.

We would find none of these objectionable.

7.13 CONCLUSION

Nowhere does the contention that manufacturing strategies should be built around the inter-relationship of product markets, technologies, work organization, etc. seem more appropriate than in considering pay. If British industry does not have a pay problem, but a rewards-and-status problem, then considering how to solve a pay problem alone could be counter productive, if not disastrous.

Acres of printed pages have attempted to promote better pay incentives and pay schemes. Yet the problem remains and is made worse as the proportion of people employed in manufacturing industry, which has a high export revenue, compared with the rest of the economy, lessens. The impact of NUPE, NALGO or even the Fire Brigades Union and their pay settlements hits home in every exporting manufacturing organization.

Changing a pay system and allying it with changes in work organization is a major test for the competence of senior management and trade unions alike. The outcome may not be fully known for perhaps two or three years. Getting it right will produce a unified and successful company. Getting it wrong could cause its demise. Endless discussion, communication and care are needed.

CONTENTION

(1) If the workforce of an exporting company wants to keep or improve their standard of living when product price rises are minimal, one important answer has to be a rise in productivity. Otherwise the company will eventually go out of business. Saying this as often as possible to the unions and to workers generally must be a key requirement of management.

(2) It is always possible that trade unions will take action or behave in ways deleterious to the long-term survival of the company. Blame for company failure must then be theirs. However, if time is spent in building up trust and an understanding of

company affairs, it seems probable that trade unions, within limits, will take a responsible attitude over pay.
(3) 'Culture' plays an important part in determining what people believe they deserve in pay. Relative deprivation is another potent factor.
(4) Pay should gradually relate to the profit/contribution/added value/revenue which product lines/sites/companies achieve.
(5) Pay-for-effort-only is totally outmoded and not appropriate to 1990s manufacturing strategy.
(6) Trade unions should be actively involved in formulating and introducing new patterns of work and payment.

A suitable set of strategies

We will continue to introduce more team incentives, with money paid for producing good, sellable products.
(2) We will push ahead with the integrated grade and pay structure covering all employees below management. We will need to introduce training and a culture which will encourage employees to learn new skills and be flexible.
(3) We will introduce unified terms and condition for all employees regardless of union.
(4) We will continue to restrict the use of Taylor type 60/80 work-study-based incentives which are tedious to introduce and because of the rating element rarely give an accurate measurement of performance.
(5) We will also continue to introduce production equipment which is more dominant over production personnel in achieving performance standards, than has been the case so far.
(6) We will end pay anomalies which stifle increased performance and, hopefully, also end relative deprivation.
(7) We will move away from the position where every last change has to be sold to the unions, to one where shop stewards actively help to provide needed changes.

8

Formulating a manufacturing strategy for the nineties

8.1 OVERVIEW

In 1985, Ferodo personnel were seconded to resuscitate BERAL, a friction material supplier in West Germany. This company had recently gone bankrupt and needed extra management to ensure that it did not go out of business completely. BERAL management, like those in many companies, had failed to generate a coherent, all-embracing manufacturing strategy on the lines outlined in this book. Improvements had been tackled on a piecemeal basis with outstandingly poor results. For example, much time and money had been spent on looking at shop-floor data collection, which was in no way a key factor in putting the company to rights. Products were being made with a sound technical reputation, but at a very high cost.

Within six months of the inception of a manufacturing framework analysis and plan, major improvements had been obtained.

Compare this West German company's condition with that of Ferodo. In March 1989, stockbrokers Hoare Govett Investment Research Limited produced a lengthy and detailed report on T&N plc, an extract from which said this about Ferodo:

> The company's continued success is built on manufacturing excellence, technological expertise and product development with the world's largest test house for friction materials.
>
> In recent years, management has accomplished an aggressive upgrading of Ferodo's production technical systems and labour strategies with a high level of capital investment, all designed to ensure that Ferodo retained its competitive edge in the longer term. The workforce and unions are playing their part in adapting

to a changing environment. Following months of negotiating, management and unions at the Chapel-en-le-Frith plant recently reached agreement in principle on very wide-spread changes in working practices at the company's Derbyshire headquarters.

This kind of comment from outsiders with proven analytical ability offset some of our frustration in not getting things done fast enough. Yet the comparison between BERAL and Ferodo shows, too, that running manufacturing companies can be as difficult in Germany as in the UK. What both companies needed was the rigorous application of the manufacturing framework and a derived manufacturing strategy.

The total process is shown in Fig. 8.1.

BERAL had a surfeit of engineers, but few people who had the business skills that a sound training in finance might yield. Complete faith in engineering, then, can lead to a disaster equally as great as one in a company where there is no respect for engineering and little or no investment. In manufacturing strategy, what is needed is a marriage of business with engineering skills. Some UK companies (BTR and Hanson) have apparently done well largely by business skills alone.

At the end of the preceding Chapters 2–7, a strategy has been declared, based on conditions and experience at the end of the eighties. Are these strategies right for the nineties? They should be tested against the 'state of the art' in manufacturing and the economic and social conditions likely to be met in the next ten years or so. This chapter suggests how it might be done.

8.2 CULTURE AND DEVELOPING A MANUFACTURING STRATEGY

While the framework recorded in this book provides a realistic basis for establishing an appropriate manufacturing strategy for the nineties, what it cannot do is lay down priorities and timescales for the achievement of the strategy.

Companies, like nations and perhaps people, are the prisoners of their own history. It is this fact and the culture which is derived from it which can bedevil the best of intentions. We defined culture as the totality of the value systems and associated decision-making process which the company had, over the years, adopted.

The nineties demand an open style of management where, as far as possible, demarcation has ended and trust is engendered. How to achieve this?

(a) The role of the chief executive or the most senior manager on site is crucial. He/she must believe in the necessity to change the culture of the company and act accordingly. He must gain trust through showing by his actions that he believes in an 'open society'. He must occupy the moral high ground, where recalcitrant members of the company cannot abuse him for continuing outmoded privileges. He must be in the vanguard of change, pushing his subordinates into what might be reluctant acceptance of loss of privileges. His own privileges should be the first to go.

(b) Education and training on a scale probably never achieved before is needed. Both ICI and Ford have now (1990–91) produced major educational programmes, especially for their manual workers. We trained widely in financial awareness and

Culture and developing a manufacturing strategy

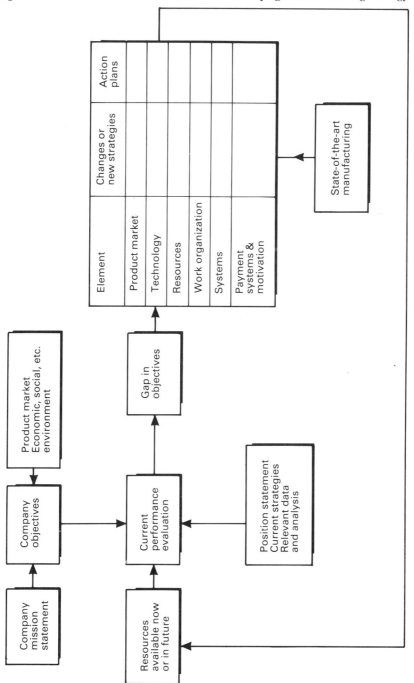

Fig. 8.1. Devising a manufacturing strategy—the process.

profit planning, in MRP II and total quality management. We used the cash we generated in the eighties for this purpose. The programme extended rather than declined as the decade went on.

More academic courses (say, foreign languages) might have been tried. As we exported so much, we needed a language laboratory.

Manufacturing companies face something of a paradox. They need highly qualified, well-motivated people, but few graduates see good employment prospects in manufacturing, especially in medium-sized companies.

The people who now work in manufacturing are often there because when they were recruited they had nowhere else to go. Yet morale will suffer if they believe that they are not being offered training which could fit them for promotion. Obviously, not everyone is in this mould, but many are.

Someone, and unfortunately it falls to manufacturing, must rectify the poor education given by an inadequate educational system. Everyone should at least be offered the chance to make better use of the talents they have. The harsh reality of training someone who will gain better employment elsewhere as a result has to be faced.

(c) In the nineties, even more that in the eighties, a company has to run as an open society. Information about the company, its results, its plans, its problems, its requirements to survive and prosper, needs to be set out regularly and as often as possible.

The profit plan, for example, needs to be discussed as widely as possible, especially with shop stewards and first-line supervision. Gone should be the days when the profit plan was a fairly esoteric document, largely prepared by the accountants. If world-class manufacturers are good at planning, then planning must inevitably be spread everywhere and involve just about everyone.

The information we gave out was never abused. It was perhaps a pity we did not give out even more.

The person who understands the company and its problems is far more likely to be sympathetic towards it, than somebody who knows nothing. If, because of a recession or a downturn in business, redundancy might be necessary, it should be as obvious to the newest recruit as to the Board of Directors. Perhaps then some kind of response which will minimize redundancies may be possible.

(d) All possible hindrances to breaking the 'us and them' situation should be tackled. Especially, relative deprivation needs to be met head on. As far as possible, conditions of all kinds should be harmonized—'staff status' should be given to everyone. (This might be abused and needs careful handling.) People should be paid in the same way. Some part of their pay should be linked to the success of the organization as a whole.

(e) Senior managers should talk to shop stewards and shop-floor operatives as frequently as possible. It is all too easy to sit in an office and keep very busy making plans, solving problems and ensuring performance is up to scratch. It might seem a waste of the 'time management' training that senior managers do; but walking about, being known, talking to people about their jobs and problems is very important in building up a relevant culture. Even the most militant shop steward tends to back off if senior management talks to him on site, where some of the problems he raises actually occur. Recalcitrance seems to thrive best when groups of teams see no one from management except their team leader and have no one else to argue with.

(f) In most factories, cultural hang-ups are probably strongest at first-line supervision level. In Ferodo, training, re-training and even more training of supervisors was introduced, along with some quite severe reaction when they did not achieve clear-cut objectives. Slowly their influence on the local culture improved. Their better outlook reflected onto those of their operatives.

The process needs to continue, harder and faster, in the nineties.

(g) Undoubtedly, as stated in (d), how people are paid is a vital component in creating and maintaining a culture and hindering its change. Payment by results schemes have probably done more to hinder change and inhibit the requisite culture in UK factories than any other factor. If operatives are treated and paid just as economic animals, why wonder when they behave solely in this way? Reward for effort alone should be totally obsolete.

(h) Union problems, too, have to be met head on. On sites where there is multi-union representation, lingering antagonism can be a deadly disease. As far as possible they should be treated equally—in discussions, in working practices and conditions, in multi-skilling, in general training and re-training. Harmonization of conditions may not be achieved easily if one union group has already a higher status than another. Differentials can be a curse. There is no way out but to abandon them.

(i) Identify and treat as special cases, people (who will be mainly managers) who have no desire to see privileges equalized and culture improved. There will always be some who view 'the workers' as an exploitable commodity and undeserving of anything more than average pay. The old class system dies hard. Again, it will be up to the Finance Director to give a lead in throwing out the gin and orange he normally has. It is unfortunate that the UK tax system has generated a series of privileges, like company cars, which are an alternative to taxable pay. They need to be abandoned and pay only given. This is less conspicuous and will normally not be detrimental to culture as company cars can be.

(j) End organizational sclerosis. Functionally split organizations can hinder rather than promote change and its associated improvements in culture.

'Staff' functions in particular seem very reluctant to see their empires fade away. A good policy in creating a different culture might be to 'give manufacturing back to manufacturing management'.

(k) Debate will go on as to whether introducing Just-in-Time or Total Quality Management or some other philosophy or technique can change the whole value systems structure in a company. It has often been tested before. Once it was computers, or management by objectives, that would achieve the appropriate end. They never did. Some changes occurred, but rarely in the depth and breadth needed. Introducing considerable investment on the shop floor might make some change, but unless operatives are prepared for flexible manning, etc., it will scarcely ever prove to be as good an investment as it should be.

From experience, everything points to a coherent, holistic approach to changing culture. Only a multiplicity of activities will convince everyone that culture change is inevitable. Production managers should not, however, underestimate the problem of achieving a social and technical relationship which is hardly ever found in the world outside the factory gates. Yet without it, the whole process of developing a satisfactory manufacturing strategy might founder.

8.3 COMPANY AIMS AND OBJECTIVES

The exceedingly rare books and articles that have been published on state-of-the-art manufacturing seldom give the performance results that keen observers would like to see. Comments on reducing stock values by a half, or improving customer service by 100%, cut no ice in companies where such measurements are viewed with disdain, if the overall financial performance of the company does not improve. The advocates of computer integrated manufacture must produce performance data that counts in the City, or in the bank lending money for investment.

By the end of the Eighties, the Ferodo Chapel results stood comparison with most of British industry and certainly with much in Germany also. There were some manufacturing units that were doing better, but probably as many as 90% were worse.

But that was the eighties. What about the nineties?

Contrary to much writing and discussion, all the evidence from users in manufacturing suggests that there is a move away from a belief that being at the leading edge of technology is essential. The focus is now on technology which may not be 'state of the art', but which integrates well with the competence and environment of the company. Technology must be subordinate to the establishment of a manufacturing framework.

What are the elements of the nineties plan? Certainly these must be the core of the activity:

8.4 PLAN ELEMENTS FOR A COMPANY IN THE NINETIES

Product market

Major elements	Subelements	Avoid/eliminate
(1) Strategic fit	(a) Product/market identification (b) Competition—knowledge updated daily (c) Strategic fit requirements (d) Globalization—must operate on a transnational basis (e) Flexibility/re-focusing (f) Quality—of supreme	— Concentration on price and delivery of goods alone — Production of standardized product — Belief that marketing is only the problem of marketing/sales personnel; it is everyone's problem

Sec. 8.4]   Plan elements   221

		importance (g) Supply what the market needs	especially that of production management
(2)	Niche marketing	(a) Identification of product/ market slots (b) Define markets which may not previously have been targeted	— Poor industrial relations, equipment and organization which hinders
(3)	Individual- ization	Product must equal exact requirement of customer	achievement of strategic fit.
(4)	Long-term supplier— customer relationship	Pricing and quality plus stability in supply	
(5)	Product design and reliability	Product design and manufacturing technology linked. Design equal at least to world best	— Lack of flexibility to respond to changing circumstances
(6)	Economic/ technological product market trends	Ability to forecast accurately	— A belief that change will be slow
(7)	Quality	(a) Total quality management— everyone involved (b) SPC/monitoring/procedures (c) FMEAs (d) Achieve BS 5750/ISO 9000 (e) Train everyone (f) Quality from raw material and component suppliers (g) Cost of quality (h) Zero-defects	— 100% inspection — A view that quality is only the responsibility of a few people — The big quality departments (as per Juran[†])
(8)	Speed of response	Response might have to be weeks, rather than months or years	— Any idea that customers will wait
(9)	Innovation	'Extra' money put into R&D to more nearly equal competition	
(10)	Packaging		
(11)	1992	Know and introduce all elements	

[†] See Chapter 2, section 2.2.12.5.

Technology

Major elements	Subelements	Avoid/eliminate
(1) World-class Manufacturer	(a) Flexibility in technology (b) Total preoccupation with quality (c) Commitment to success (d) Strategic vision	— Attitudes like 'we only serve home market'
(2) CIM	(a) Institute proven technology within a long-term plan, with discrete segments (b) CAD/CAM Computer-aided engineering Flexible machinery centres/ Flexible assembly centres market requirement, especially in improving quality (d) Continuous flow manufacturing	— Unproven technology — Attempting too much at a time — Views that major levels of stock are necessary and long lead times are constant
(3) Flexible manufacturing system (4) Advanced manufacturing technology (5) Robotic cell manufacturing		— Training which is haphazard and not demanding
(6) Optimized product technology (7) Group technology	(a) Always look at key operations, not all (b) All operations should provide saleable value Multi-skilling necessary	— Waiting time, intermediate inspection, storage time.
(8) Vision systems in quality Assurance		
(9) Links with MRP II	Integration is vital between design, planning and production	— Old-fashioned functional organizations.

(10) Automated materials handling and storage
    (a) Robots
    (b) Automated guided vehicles

(11) Plant layout to minimize space and support new technology

(12) Short-cycle manufacturing

Resources and cost control

Major elements	Subelements	Avoid/eliminate
(1) Contain working capital. Minimize at all levels and activities	Stock control Debtor control Creditor control Cash management Cash forecasting	— Bottom-up inventory control — non-targeting
(2) Financial/cost control	Daily performance reporting	
(3) Material utilization	Technical and operational analysis. Relationship with quality	
(4) Relationship of costs to benefits achieved	Top-down planning of resource allocation Buy in resources when necessary Privatization	— Belief that everything needs to be done in-house
(5) Labour performance improvements	Performance appraisal related to pay	
(6) Research and Development	Strategic spend	
(7) Distribution		

Purchasing

Major elements	Subelements	Avoid/eliminate
(1) Integration of suppliers into order processing chain	Suppliers to conform to company's own quality standards JIT Long-term relationship	— Confrontational approach with price reduction or minimum price rises paramount
(2) Strategic approach to purchasing	Value analysis Integrate MRP II systems	
(3) Reduced number of suppliers		

Work organization and training

Major elements	Subelements	Avoid/eliminate
(1) Work organization related to achieving self-worth	(a) Focused businesses (b) Focused factories (c) Focused product lines	— Functional hierarchical organizations which restrict change and add cost to the organization without giving much benefit
(2) Team activities Build up team competence to carry out needed activities	(a) Team Activity covering • production • maintenance • Health & Safety, etc. (b) Paid on an input/output basis	
(3) Supervision—a separate development	Training/motivation/payment	— Status differentials which generate class divisions
(4) Communications	Debate with whole workforce on profit plan and performance	
(5) Consultation	Second-tier Boards as in West Germany	— Payment systems which inhibit setting up teams and new form of work organization
(6) Training—to achieve cohesion at all levels and create requisite skills	Strategy to provide • technical/management skills • cohesion • organizational changes • organizational understanding	— Training without a

Sec. 8.4] Plan elements 225

		strategic purpose
and motivation to achieve 1990s style strategy		
(7) Multi-skilling	(a) General skills acquisition (b) Problem-solving skills to create employee flexibility of a high order	
(8) Introduce 'Social Charter'		

Information Technology systems

Major elements	Subelements	Avoid/eliminate
(1) Information technology in • Text message handling • Business graphics • Electronic mail etc.	(a) Local area networks/ communications architecture/ wide area networks (b) OSI/MAP/TOP protocol (c) Intelligent workstations (d) Company-wide information exchange (e) Shop-floor data collection (bar coding, etc.) (f) Programmable logic controller (g) Data management system (h) Design support systems (i) Business control systems (j) EDI	– Non-interactive terminals
(2) Expert systems	ES to cover a wide variety of planning and tooling requirements	
(3) Simulation	Simulate material flow and manning	
(4) Systems MRP II (Class A user)	(a) Database/bill of materials (b) Material requirements planning (c) Capacity requirements planning Rough cut scheduling (e) Shop scheduling (f) Performance monitoring	– Systems which do not recognize actual or potential bottlenecks
(5) Management Accounting	(a) Top-down planning (b) Manufacturing Accounting and	– Bottom-up business planning and

| and Business Planning | production information control systems
(c) Links within CIM essential | management accounting |

Motivation and payment systems

Major element	Subelements	Avoid/Eliminate
(1) Generation of a self-worth culture	(a) Multi-skilling (b) Communications (c) Job enrichment (d) Payment of teams on an input/output basis (e) Share options (f) Social profit sharing	– Hierarchies and status barriers
(2) Team working		
(3) Organizational changes which allow greater involvement in decision-making at employee level		
(4) Integrated pay	(a) Performance-only pay systems a possibility (b) Performance appraisal allowing differential payments to be made, especially to those with greater skills (c) Communications on company performance	– Any idea of paying people purely on simple work output
(5) Harmonization of conditions		
(6) Skills acquisition related to pay		
(7) Gainsharing or company-wide performance-related pay		
(8) Added Value earned systems determining productivity and pay		

8.5 WHAT KIND OF A COMPANY MUST WE BE IN THE NINETIES?

The record of what kind of company we needed to be, as set out in Chapter 1, might have served well in the 1980s. What changes, if any, are needed for the 1990s? If any changes are needed, will these be fundamental or merely marginal?

Is it absolutely necessary to become a 'word-class manufacturer'? Even if it is not completely desirable, then the requirements of such an organization go a long way towards helping any manufacturing organization to be successful.

Culture should play a large part in determining what kind of company we should be. For most British companies, creating an appropriate culture is very hard. They have had generations of class conditioning, which no one can throw off easily. What culture is needed and how it might be achieved has already been discussed, but the difficulties remain.

Responsiveness is now probably more important than it was. Competitive challenges can be swift and immediate. They can be ruthless in attacking products and markets. On some occasions it might be necessary to re-write the profit plan when it has only run for six months or so.

Knowledge of product markets and associated competition probably has to be updated daily. Response times need to be cut to the bone. This also means that manufacturing needs to be responsive to changes in demand, product type and quality, in days rather than weeks.

Towards the end of the eighties, Ferodo began to invest in manufacturing capacity beyond that currently needed by the sales forecast. Original equipment orders for motor components tend to come in discrete chunks. An inability to make such a 'discrete chunk' loses the business, perhaps for a long time. This throws a burden on investment procedures and local work organization.

The need for immediate response suggests that an absolutely rigid strategy goes out of the window. Quickness on the feet demands flexibility at all levels, especially in thinking. Organizations need to be in place which promote flexibility and speed of response. The old functional breakdowns are no longer valid.

How does all this re-shape the 'kind of company we want to be'? For the nineties, this is the likely answer:

(a) Profit and cash
The final judgement on the success of a company must still be the profit it makes and the cash it generates. Anything less should be unacceptable. While it may not be necessary to make profit at a requisite level immediately, it should be generated sooner rather than later. Return on capital is a useful measurement of effectiveness and efficiency. Payroll as a percentage of added value is still one of the best indicators of productivity. It is a figure which needs to be improved year by year.

The manufacturing companies in the UK which have concentrated on getting their financial results right are rarely the ones that get taken over or made into mincemeat by foreign competition.

When T&N made a bid for the Associated Engineering Co., the business press was vocal in their support of A.E. Here was a company that had invested heavily in high technology. It was just unfortunate that it had failed to achieve reasonable profitability. It fell to a more financially hungry company.

It is not without reason that the companies in the UK that have not been upset by foreign competition are those which put profit and cash high on their list of objectives. Those that did not, like Rolls Royce when they were developing the RB211 engine, succumb. The conclusion is obvious.

(b) People
People should really be linked with the culture of the company. The one makes the other. The culture of the company needs to be based on an open society with harmonization of conditions. As far as possible, relative deprivation should be eroded and the difference in status amongst all employees—management, supervision, operatives alike—should be eliminated.

Training of people should not be something that is switched on and off just because profit appears to be lessening and costs need to be cut. British companies trained too little in the eighties. The possibility is that they will train too little in the nineties. Once training is adequate, work organization is crucial in ensuring that people are both motivated and will give something akin to their best to the organization. Team organization appears to be the best structure for providing organizational effectiveness.

Perhaps, cruelly, the numbers of people in manufacturing organizations will inevitably have to decline. By the end of the eighties, Ferodo had doubled its sales, using 35% fewer people than they had in 1980. This is a trend that must continue. The degree of market expansion available will rarely provide the opportunity to ensure continual rises in productivity using the same (or perhaps even a higher) number of people.

(c) Communications and people are important
The Japanese have two relevant views on communication.

(i) They communicate prior to (or perhaps after) the end of each shift. There are few monthly meetings. They do not feel they are necessary.
(ii) They keep the message simple. They have found that their employees are most concerned about their local jobs, what can be done to improve them, what problems they have had and are likely to have. If the outside world is mentioned at all, it is only in relationship with them and how it will affect them.

The key member in the communication chain is the first-line supervisor. If he communicates daily in a logical way which supports senior management, all may be well. If not, then any monthly scheduled meetings could be a waste of time.

Communications have never been so important. Getting the message across should be of major importance.

If everyone's pay and bonus depend upon the company performance, then the reasons why the company is doing well or badly need to be totally explicit.

Communications—between managers and employees and back the other way—can serve to transmit information and ideas about improving job and company performance.

(d) Recruitment

Recruitment should be done judiciously. Early retirement is perhaps the pattern.

There are inevitably going to be disappointments in equalizing privileges and status. While self-discipline is an ideal, it should always be capable of being re-inforced by organizational strictures.

One way of beating the demographic trend is actually to call on fewer people to man machines and work on the shop floor. If reducing numbers in total is inevitable, reducing numbers on the shop floor is paramount.

In the end, a manufacturing company should be a classless society if there is to be any hope of equalling Japanese and other Far Eastern competition.

(e) Response and innovation

If response to competitive challenge has to be swift and immediate, why stop at the marketing activity? Everyone needs to respond quickly to every activity—stretching from the time taken for an order to be put into the MRP II system, to how long a product stays in the warehouse until it is despatched.

The general pattern in British industry has been one of comparatively slow adaptation to change—if adaptation occurred at all. Everyone used to take their time about doing anything. The profit and loss account was produced ten or maybe more days from the end of the month it was reporting on. Hour-by-hour control of the shop floor was not part of the control function. It took anything up to a week to get an acknowledgement to a customer.

All this is nonsense. Once, in Ferodo, under a clerical work control system, all work was cleared up on the day it arrived. When the control system was relaxed, delays began to occur.

Just-in-Time is a philosophy which applies everywhere—office, laboratory, computer room, maintenance and the shop floor. If it seems necessary to introduce control mechanisms to achieve JIT in the office, then so be it.

With responsiveness must still go innovation. Japanese factories such as NISSAN actively encourage the workforce to get together to discuss profitable change. If it can be done in Sunderland, why not elsewhere?

(f) Customers and sales

It is likely that responses to customer demand should be made daily, if not hourly. The need to relate products and customers on perhaps a one-to-one basis has already been mentioned. Customers always were important; they are even more important if they can get the same goods that they are currently buying in the UK, from Japan at 10% less price.

It seems ridiculous to send salesmen to Germany when their German does not even stretch to 'Wie geht es Ihnen?' Nor is it reasonable to have telephone receptionists who only talk in the broad local dialect.

Selling is everyone's job, especially that of production managers. It is too important to be left to salesmen alone. Everyone should be involved in achieving strategic fit—gaining manufacturing flexibility in quality, service, cost and product design. Everyone should be aware of why it is not being achieved (if it isn't) and what is being done to achieve it. Achieving strategic fit should be a useful discussion point when managers and shop stewards meet. There should be a constant debate about it throughout the company.

(g) Quality
It should be a matter of profound shame when rejects occur, when a mistake is made in entering a new order, or when a lorry is misloaded. Like just-in-time, 'quality' should be all-pervading and applicable right across the company.

Sadly it may be necessary to introduce disciplinary procedures to ensure that quality is high on everyone's job objectives. Making mistakes could be human, but forgiving them when they cost £50 000 or more is something that needs to be thought about and perhaps discipline imposed.

Believing that quality applies just to products leaving the company and nothing else will do nothing to engender a 'quality ethos'. The whole tenor of making mistakes and their aftermath should be raised. Getting things right—all the time—must be a key management task, with making a mistake regarded as anathema.

(h) Importance of the shop floor
Hopefully there are now few manufacturing companies where directors do not visit the shop floor regularly, where they are not known or spoken to. It was not always so.

The shop floor can only grow in importance as production managers take part in discussions on strategic fit. If the front line of customer requirements now starts on the shop floor, then equally the shop floor must go out to the customer. Customers, if they are to form a long-term relationship which will guarantee business for a number of years, have to have faith, not in the salesmen buying them an expensive lunch, but in the production manager who can produce his goods to a quality, design and flexibility which probably has never before been achieved.

The need for outstanding talent on the shop floor has never been greater.

(i) The technological base of the company
Of course the technological base of the company still needs to be advanced, perhaps considerably. At the end of the eighties, in Ferodo, there were still far too many jobs on the shop floor which were boring to the point of being mind-distorting. These should be eliminated. How far, though, should the real world of manufacturing link up to a Science Policy Research Unit paper (published by Sussex University)? This states: 'Britain's technological deficiencies are long standing and well known. There are no convincing signs in the 1990s of any change for the better in the historical trend'.

A major problem in the UK, it is said, is the short-termism of manufacturing industry towards investment. Other nations, which are truly innovative, take a long-term perspective and have accounting procedures which allow investment to be introduced without an immediate, satisfactory return.

It is no part of a product manager's job to rectify the misuse of national resources over many decades. He must gain strategic fit. He must help his company be competitive. His investment must give an adequate return. Advance the technological base of the company by all means, but not because some academics believe it to be necessary. Certainly, new technology is not an alternative to gaining relationships throughout the company which allow everyone—within limits—to give of their best to the organization. Both are needed.

(j) Resource usage

During the downturn in business evident in 1990 and 1991 in the UK, there were numerous reports of companies going out of business because of their high borrowings. Some manufacturing companies had invested wisely, as they thought, in super technology only to find they could not afford it when business declined. The companies that resist downturn on business best are those with strong balance sheets. This in large part reflects the resources collected by the company and how they are used. High debt equity ratios are not useful if they exist for any length of time.

Working capital should be reduced to a minimum. While perhaps JIT and Total Quality Management will help to reduce stocks and improve material utilization, reducing working capital as a whole needs a separate strategy. Cash generation should be crucial, especially in providing money for investment.

From time to time, this book has considered whether there is any one idea, philosophy or activity which will push a company into being in the front rank of its type. If there is one factor that meets this requirement, it is *the acquisition and use of resources in creating profit*. The relevance of CIM or JIT can be more clearly seen in relating resources to profit than probably in any other way.

(k) Rewards and punishments

If one thing helped to destroy the commercial ethic in the UK in the eighties, it was the educationalists' belief that schools should no longer test their pupils regularly, and indeed that team games where winning and losing played a part should be abandoned. In a world where organizations competing on a world-wide basis are quickly made aware of whether they are winning or losing, internal rewards and punishments are vital.

Managers in manufacturing may be running contrary to the latest educational convention when they insist that organizational success and failure should be strongly related to what people at all levels earn. Failure should bring not just loss of pay but, perhaps, loss of position as well; but pay changes should probably dominate.

As world-wide competition increases, then so too should internal rewards and punishments. Probably the reward that should be the most inadmissable is that, having failed, a manager should receive a major pay-off when he leaves the company.

Competition can bring cruel results. Manufacturing companies might need to create a reward-and-punishment ethos, contrary to that in existence in society as a whole.

(l) The 'green' revolution

Throughout the eighties, the pressures for a cleaner environment grew. They will grow further and faster in the nineties. Ferodo, because it used asbestos as a raw material for many decades, was always very careful about pollution and handling waste material. Despite surveys that proved that asbestos had been handled as safely as possible ('Ferodo Mortality Study'. M. L. Newhouse and K. R. Sullivan. *British Journal of Industrial Medicine* 1989. **46** 176–179) asbestos was phased out as quickly as possible. The pressures to abandon the use of the material were too great.

Such 'green' pressure can only increase. Most factories use quite dangerous chemicals and perhaps processes. Their control in use and, above all, in waste disposal will become a bigger and bigger burden on production management. They should accept it very willingly.

Perhaps one of the most beneficial aspects of maximum material utilization is the minimization of waste and its associated disposal problems.

(m) Management competence
Most of the preceding items would not really have been out of place in the mid-eighties. The most important change has been kept until last.

Management competence in the UK certainly seemed to improve in the eighties. However, there is a view that much of the productivity improvements gained then were largely due to a once-and-for-all reduction in people which occurred during the recession of 1980–83. Once business picked up, and numbers of people employed remained the same, productivity was bound to rise. Add to this an improvement in investment, and profit increases were nearly inevitable.

The rest of the world is not standing still and the legacy of past redundancies—even ones made in 1990–91—will soon disappear. The degree of management competence needed to be successful in the nineties is a quantum leap higher than that required even 6 or 7 years ago. When seen against the general competence of local government or perhaps some service industries, the differential between their productivity and that required by manufacturing industry needs to be immense.

In every way—in reducing working capital, in raising productivity on a year-by-year basis, in understanding and responding to markets, in the slickness and the quality of every activity being carried out—manufacturing needs leadership and expertise, which not every company has yet recognized.

Training will help, but alone will not solve the problem. Only recruiting the best people available might do so.

8.6 KEY QUESTIONS

Most reasonably competent managers will be able to draw up a list of questions which need to be asked if a satisfactory strategy is to be determined.

The questions we asked ourselves and the points we thought needed to be classified are listed in these seven subsections below.

8.6.1 General

(1) Review performance achievements in the last 5 years, especially return on investment, operating profit/sales, added value earned.
(2) Review the number of employees and their pay set against revenue and profit earned—by total company, product line, department and function.

(3) List current products made and their position on the product life-cycle.
(4) List current customers and indicate why they buy from us.
(5) List all actual and potential environmental influences—health and safety, etc.
(6) Consider the competence of management and especially the Chief Executive in:

- making and achieving plans
- understanding finance
- understanding the business
- appreciating what must be done to improve company performance.

(7) Review the company organization structure and determine how it needs to be changed (if at all) to ensure survival in the nineties.
(8) Does the company have adequate leadership, communications, culture, ability to change.
(9) Determine the technical competence of management (as opposed to management expertise).
(10) How do we create change? Test out all the changes which appear to be necessary under the headings listed in the standard approach to making a manufacturing strategy. Ensure that all relationships are fully worked out. In particular determine how:

- the implementation of MRP II
- a new payment system based on team incentives,
- robotic cell manufacture

can be handled and record the likely effect each will have on the others.

8.6.2 Product market

(1) Define competitiveness.
(2) Review all product markets using the profiling technique.
(3) Determine what has been happening in the last 5 years in:

- revenue earned
- profit/contribution earned
- prices
- market share.

(4) What growth is expected in product markets in the next 5 years?
(5) What market share of what product market seems possible? How do we achieve this? List:

- new technology
- lower costs/lower prices
- improved product range
- quality
- service/stockholding

(6) What new competitors might we have to face which are not currently attacking us?
(7) Report on all current competitors. How can we improve links with customers and potential customers?
(8) What pricing strategy has been followed? Has it worked?
(9) What new relationships with non-UK manufacturers can we develop (Joint Ventures)?
(10) How do we go about improving our quality? What are current quality standards? Are they good enough?
(11) What strategic fit can we determine?

8.6.3 Technology

(1) Become fully aware of the current trends in technology. Visit suppliers and if possible gain access to those organizations which have installed new technology—robots, CIM, FMS, etc. Consider carefully whether any application is readily transferable.
(2) Determine where and what new technology might be deployed on the factory. Review all plant in use, its age, its efficiency, its utilization. Especially look at the following:

- plant layout (how much it could be improved)
- efficiency (cost, book value, and labour needs, of equipment)
- use of computers in process control and planning (MRP II)
- current skills of the workforce
- whether manufacturing systems give good control

(3) What do customers want in the way of:

- new materials
- better service
- broader/better product range?

How far does current technology hinder strategic fit?

(4) What are our competitors doing? Have their technological changes been a challenge to us?
(5) What is the money situation? How much will be needed? Where will it come from?
(6) What changes in Technology will have an effect on the work force:

- its competence
- skill and training
- work organisation?

(7) What methods of financial evaluation and control will we use? Are these good enough?
(8) The future appears to lie in integrated computer systems. Does our plan reflect this?
(9) What can we do to minimize the risks of applying new technology?—step-by-step, buying known and proven hardware, etc.

8.6.4 Resource utilization

(1) List all the key resources we have used over the last 5 years and relate them to achievements in revenue, operating profit, added value, etc.:

 - people by function, department, status/grade
 - material by product line
 - energy
 - fixed factory cost
 - SD and A

What anomalies can be seen? Are these explicable? If not, they should be challenged.

(2) If people are the most expensive resource, how do we contain or even reduce their cost? By:

 - re-organization
 - work simplification
 - privatizing those functions and activities which lend themselves to this practice.

(3) How do we make sure that everyone has a proper job to do for at least 95% of the time?
(4) Material is probably the next most important cost. What is the current yield and how can it be improved?
(5) Review the purchasing function and determine whether it can be made more effective.
(6) Are cost controls adequate? Do we carry out top-down planning as well as we should?

8.6.5 Work organization and training

(1) Consider the current work organization and determine how far it will help or hinder the improvement of the following:

 - productivity on a year-by-year basis
 - quality of our products
 - service/flexibility needed to improve our revenue
 - reductions in working capital

(2) What key management grades need to be made more effective?:

 - all management
 - senior management
 - first-line supervision
 - technical personnel.

How do we do this?

(3) If hierarchies and line and staff relationships are obsolete, what replaces them?
(4) What is a team? How do we measure its performance? What inputs and outputs are needed?

(5) How do we get more consensus in setting corporate objectives?
(6) Why do we train? For what purpose? What type of training do we use? Is it effective?
(7) How can we achieve corporate cohesiveness? Through:

- finance training
- MRP II training
- action-centres leadership.

How else?
(8) How can management and the shop floor see eye to eye more often?
(9) How good are communications? How can they be improved?

8.6.6 IT and systems

(1) Define information technology.
(2) Define a system. Where do the two touch?
(3) Why do we need information—for what purpose?:

- resource planning
- profit planning
- monitoring progress
- determining where performance can be improved
- to say how much things cost
- to relate one action with another, like taking in an order and determining if it can be made on time
- to answer 'what if' questions.

What about speed and accuracy?
(4) Why not ask all managers and supervisors to write down their objectives and then record the information they have to help achieve them. What other information is needed apart from this? Why?
(5) Computer hardware is cheap. Software developed internally is expensive. Why do we continue to develop our own internal software?
(6) Is our computer technology/hardware obsolete?
(7) Are MRP II and Management Accounting the two systems that we should be introducing/improving at the expense of all others?
(8) Do we have an appropriate database—in terms of information structure, and accuracy?
(9) Who in the company is capable of writing a user Specification for either MRP II or Management Accounting?

8.6.7 Motivation and payment systems

(1) What has been happening to pay over the last 5 years by:

- union
- function

- direct/indirect workers
- staff/management.

when compared with:

- RPI
- profit
- revenue
- added value
- trends in other similar companies?

What does the result suggest? What corrective action of the current payment systems is needed?
(2) Consider the strengths and weaknesses. Do they support required changes in:
   - technology
   - work organization
   - resource utilization?

If not, what are the alternatives?
(3) How are the proposed alternatives to be sold to the unions?
(4) How quickly can we get the unions to help in devising new payment systems for their members?
(5) Have we any 'red circle' jobs, where the pay being given is way beyond what any reasonable system should produce? If so, how are they to be tackled?
(6) Is harmonization of conditions a vital component in any new payment system?

8.7 STATE OF THE ART

Most competent managers will be aware of current trends in technology, systems or methods of payment. Consultants, academics, magazines will only be too ready to put forward what is considered to be the last word in producing a manufacturing strategy. Always be wary of the latest piece of wizardry from the local Business School.

Ferodo with its world-wide links, especially in Japan, Germany and the USA, could debate carefully the important changes which were occurring on the road to becoming a world-class manufacturer. Even so, a state-of-the-art review is important. Managers should consider this, and determine which changes are of crucial importance and which not, in developing a manufacturing strategy. A grading system might be used:

A — Of vital importance and immediate concern; without this we may not survive.
B — Very important and needs to be done as soon as possible.
C — Important, but can wait—a little.
D — Does not seem too important to us, despite what consultants and others say.

Gaining money for investment in new technology is important—must it come from cash generated internally or is it so important that we have to borrow?

8.8 KEY ELEMENTS

(1) Little has been said about strengths and weaknesses, which would normally form a significant part of strategy formulation. This is mainly because they should emerge from any evaluation of using the manufacturing framework and especially strategic fit. Certainly evaluating company culture would also help significantly.

Data analysis should also indicate where the company has had success and where perhaps it has failed.

Producing a 'what kind of company should we be' statement should indicate strongly where changes are needed, whether totally or in emphasis which will hopefully eliminate weaknesses.

(2) Strategies are valueless if they do not address the most fundamental part of the business—how to gain sufficient business to help achieve satisfactory financial returns.

(3) Definition of the market is crucial. Equally, differentiating company capabilities compared with market requirements is essential. Strategic fit analysis should give the right answer, once markets have been identified in detail.

(4) Setting demanding targets. Some apparently important objectives, such as revenue earned per person, may be less important than some others, such as operating profit earned per person, or added value earned as a ratio of pay roll.

(5) Many Japanese companies, faced with competition from South Korea and Taiwan, are retreating further into high added-value products, often at the luxury end of the product market spectrum. In achieving high added value, design is often important along with technological flair.

(6) Some years ago, when corporate strategy was in vogue, 'growth' and how it could be achieved was always a major point for discussion. For many companies, 'growth' has meant major investments, extra labour, risk in marketing. This has been as much a road to disaster as not trying to grow at all, especially with the economic rollercoaster seen in the UK.

Growth and risk go together. What risk appears acceptable (or perhaps even desirable) must form part of the discussion about strategy.

(7) Where is the money to come from? The dilemma of keeping profitability high, at the same time that major investment and other changes are being made, is real. How it is settled will influence strategy, perhaps for some years to come. Will borrowings rise significantly? How much can be generated internally?—by far the best way, but how long will this then delay needed investment?

(8) Interrelationships matter a great deal. The production framework should be used to provide an interrelated set of strategies.

(9) Option. It is possible that the use of the manufacturing framework, plus state-of-the-art data, will throw up a series of options. Which, or perhaps how many, will depend upon:

- urgency (how quickly is it necessary to carry through an option to ensure company survival?)
- degree of risk management is prepared to accept
- competence of local management

(See also section 8.10 Timescales.)

(10) In large part, the options chosen will depend upon the type of manufacture and markets being served. In Ferodo's case, for example, with major product quantities going to original equipment manufacturers (OEMS), it was important that one-to-one relationships were achieved and maintained. A production unit making domestic appliances or fashion clothes would see relationships and options differently. Cost structures will also vary depending upon the type of manufacturing undertaken and markets served. Ultimately, though, nothing should diminish the importance of profit and return on capital employed.

(11) Competitive edge. A useful option is to consider how to improve competitive edge, i.e., what must be done to enhance strategic fit. Any one of a number of items could be put forward all directed towards gaining competitive edge:
 - quality
 - responsiveness
 - low cost
 - short lead times and due date delivery
 - good design
 - reliability and delivery
 - customer response
 - the product itself
 - after-service.

(12) Unit labour costs. The recession in the UK in 1990–91 has again shown that despite major investment in the eighties, unit labour costs quickly rose as business plummeted and inflation increased. The problem of paying what can be afforded has not gone away. This is proof if it was needed; that manufacturing strategy could be as much about work organisation and payment systems as technology.

8.9 ENVIRONMENTAL ANALYSIS

The analysis necessary to carry out strategic fit evaluation will help considerably in arriving at a satisfactory environmental analysis. It will not help much in three other areas—technology, social conditions and economic trends.

(1) Technology

Many publications, from *The Economist* to *Works Management*, will from time to time give some indication of technological trends. The 'state of the art' in technology should be constantly assessed. This applies not least in information technology and systems. Consultants will be only too ready to sell extravagant solutions to comparatively minor problems. A broad knowledge of trends will avoid mistakes.

'What's happening in Japan?' might be a more fruitful question than pursuing the latest 'best factory' in the UK; but visiting other factories is essential.

(2) Social conditions

Social trends may not sound particularly important, but they are. The society which becomes progressively more ill-disciplined is not one that will provide a disciplined workforce in a factory. People in manufacturing should be aware of the social trends which could help or hinder their activities to ensure that their influence is either benign or at least offset in large degree.

An interesting document is the 'Social Charter', which emphasizes social rights within the EC. It postulates the following amongst other requirements:

- Improvements in living and working conditions
- Right to health protection and safety at the workplace
- Employment and remuneration (all occupations to be fairly remunerated)
- Right to social protection
- Right to freedom of association and collective bargaining
- Right to vocational training
- Right of men and women to equal treatment
- Right to information, consultation and participation of workers.

(3) Economic trends

Economics, the dismal science, is not held in very high esteem, yet poor economic forecasting can wreck a company which has got everything else right in terms of strategic awareness. Perhaps the saddest remarks made during the downturn in business in 1990–91 have been by those who borrowed large amounts of money, only to find that interest rates have rocketed up and sales disappeared. Investing in new technology alone was obviously not the answer to problems of survival. 'When?—' and 'how?' were the vital questions.

8.10 TIMESCALES

The pace of change will largely depend on:

(a) The previous history of the company. Its successes and failures. The relationships between managers and employees. The culture of the organization. The leadership and drive of senior management.
(b) The expertise of management who will be involved in introducing the strategy, especially new technology.
(c) The state of competition. If this is fierce and getting fiercer, then risks may have to be taken in putting a new strategy in place.
(d) Financial and other resources which can be made available. Perhaps consultants may be necessary in introducing change. They will inevitably be expensive. Are they worth their cost?
(e) Changing organizational structure may need new occupants for managerial positions. These may have to be recruited. How soon can it be done?
(f) The state of the market. What position on the business cycle has been reached? It is normally easier to make radical change on the economic upswing.

No one should ignore the length of time it might take to drag a multi-union site into strategic reality. Equally no one should ignore the pace throughout the world at which conditions are changing.

8.11 PROFIT PLANNING

Profit planning has been defined as a disciplined method whereby the environments impingeing on an organization are analysed, the available resources and internal competence identified, agreed objectives established and plans then made to achieve them. The company mission statement and associated strategies are implicit and a major part of the process.

The profit plan should be the main engine of motivation in the 'train' principle. It should establish a discipline in setting and achieving objectives to which everyone should conform.

It should help to ensure organizational cohesion. It should establish how resources are to be allocated and the relationship between resource, its use, and its planned achievement.

The plan should be all-pervasive.

Action plans should be instituted wherever a relationship has been made between a strategy and a desired result.

8.12 CONCLUSION

Fears for manufacturing in the future

This book has described with as much candour as possible how one company—Ferodo—was brought back from the brink of disaster to being a reasonably high-tech, good profit-earning company.

Some quite fundamental changes were made. In earlier years they might have been called revolutionary.

Yet, even knowing what the state of the art suggests should be in the plans for the 1990s, even knowing what related Japanese and American companies are introducing, even appreciating how good competitors are, there must still be some doubts about the future, for UK manufacturing generally.

(a) The production framework takes on enhanced appeal the more competition takes off. Somehow the inadequacies of the British educational system must be offset by enhancing the 'train' principle. Everyone should be pushed on board, irrespective of competence, and then be pulled along towards a reasonably tough goal. Rewards and punishments must be in place for this to happen.
(b) Spending on R&D has to rise, even at the expense of profit in the short-term. At the same time, money must be spent on training—just about everyone. Neither of these outlays will have a short-term pay off.
(c) Managers must help to get rid of the old class basis of running companies. With society generally class-ridden this is a formidable challenge to the upbringing and innate beliefs of many reasonably well educated managers. Unless considerably more egalitarianism is evoked, than often still appears to be present, life can become precarious.
(d) Nothing is standing still. The Japanese are not waiting quietly for the rest of the industrial world to catch up. What was useful and productive five years ago,

could now be out of date. Once, as Ferodo found, getting financial control over the company was difficult. It should no longer be so. There is no management time left to keep solving problems which should have gone away in the 1970s.

Management should concentrate on getting strategic fit, getting quality right, absolute minimum stocks, introducing work organizations and motivational procedures which make sense in the 1990s.

(e) The Japanese largely developed their successful industry by importing Western technology at great expense. No one in the West, especially in the UK, should be made to feel ashamed of importing Japanese technology in the profound hope that the result will be the same.

(f) Along with imported technologies, management should not be ashamed to use what limited skills may be available in key subjects. Developments, especially technical ones, are getting beyond many middle-aged managers. Bring in consultants to carry out major changes.

(g) Collaboration with other companies could be one way of minimizing risk in R&D and in production technology. If the two companies operate in separate and distinct product markets, no harm should be done and much benefit gained.

(h) Never be ashamed of making comparatively mid-tech products. If they are made efficiently at low cost, then it is possible for a niche to be filled and useful money made at the same time.

Japanese factories may be becoming quiet, cloistered halls of assembly, where as much thinking is done as manufacturing. It may still not be so in the West—yet. The chronic shortage of engineers in the UK in particular may preclude a rapid change from semi-smokestack industries. Somehow these must be made to serve for some years yet, to produce wealth and added value for future material investment.

(i) Everything, however, turns on having management which can readily identify future strategies and have the intelligence and dynamic to push them through. This, as stated earlier, may mean running manufacturing industry at a level of efficiency and with a management competence far higher than either local government or service industries might ever have achieved or perhaps need to achieve.

This then remains the major question. The phoenix may have arisen from the ashes. Will it continue to fly? Here lies the test for Ferodo and for all other manufacturing units in the West.

Appendix 1: Ferodo

As much of this book revolves around Ferodo, a brief description of the company might be appropriate.

Ferodo Ltd was founded by Herbert Frood in North Derbyshire at the end of the nineteenth century. From his successful attempts to improve the braking on horse-drawn vehicles crossing the Derbyshire Peak District, has grown the most comprehensive manufacture of vehicle friction materials in the world.

With its associates, it has factories in West Germany, Italy, India, Nigeria, Spain, South Africa, Zimbabwe, Zambia and the USA. It has technical agreements with Japanese friction material suppliers.

It makes brake linings, clutch facings and disc brake pads, in composite and metal materials for both the original equipment and replacement vehicle markets. It has the largest brake material testing facility in the world.

A major part of its UK production is exported. Its material technology is also exported to its world-wide associates.

The world-wide manufacturing facilities ensure that globalization is standard practice. The philosophy of the world car is also served by the widespread manufacturing activities and technical agreements.

Asbestos was once a major raw material in product formulations, but is now practically eliminated in current production.

While cars are the most obvious users of Ferodo materials, they are also stipulated for motor cycles, trucks, tractors, buses, railways, many engineering applications, oil rigs and mining activities.

Ferodo's Head Office and main UK manufacturing activity is in Chapel-en-le-Frith, Derbyshire.

Appendix 2: List of definitions and explanations of terms and acronyms used

To avoid breaking up the flow of the narrative, explanations and definitions of the terms used in the book have been kept to a minimum in the text. Many readers will already be aware of and could probably define the majority of the concepts and acronyms used, but for completeness they are now recorded briefly.

ACAS   Abbreviated form for the conciliation and arbitration services, which are government-funded and provides companies in dispute with their workforce with an appeal procedure. ACAS officials act as referees,

ACL   Action-centred leadership. A recommended process whereby supervisors and managers of all grades are taught leadership, including performing leadership tasks. Team activities are strongly stressed.

ADDED VALUE   The value added to materials and other purchased items which provides, as a result of productive activities in the firm, the sum out of which wages, salaries and administrative overhead expenses are paid, leaving the surplus as profit. (Engineering Employers' Federation definition.)

AMT   Automatic/automated manufacturing technology. Equipment which is largely planned, set and controlled by computer or process controller.

CAD   Computer-aided design. The application of computers in the design process. Much stereotyped routine design is done by computer with specialists adding limited bespoke sections of the design.

CAM   Computer-aided manufacture. The use of computers in carrying out manufacturing activities such as the programmed use of wire eroders in tool-making.

CE   Chief executive. An American expression which is used in the UK to describe the senior operating manager, a slightly different definition than that which would be used in the USA.
CELLULAR MANUFACTURE   Usually associated with robots, where an integrated manufacturing activity is established which can operate in isolation from other activities.
CFM   Continuous-flow manufacture. Perhaps the opposite of cell manufacturing. A production line is established where products are processed from start to finish, in an uninterrupted flow.
CIM   Computer-integrated manufacture. A philosophy which links islands of technology or stand-alone computers. It covers all aspects of the business, not just manufacturing.
COSHH   The Control of Substances Hazardous to Health, Regulations (1988).
CSEU   Confederation of Shipbuilding and Engineering Unions. A union group which covers skilled and semi-skilled employees engaged in tool-making, general engineering and maintenance activities.
DCF   Discounted cash flow. An accounting procedure, essential in measuring the potential of a new investment. It puts diminishing importance on foremost cash flows the further into the future they are earned. It has largely replaced the 'payback method' of investment appraisal.
EDI   Electronic data interchange. A means whereby information can be transferred between organizations.
ES   Expert systems. An information technology process, directed at minimizing local systems design.
FGS   Finished goods stock. All products which are in a condition to be sold to a customer.
FMEA   Failure mode effects analysis. An analytical procedure, within quality assurance routines, to detect why products fail to meet standard requirements.
FMS   Flexible manufacturing systems. A methodology embracing technologies and systems which enable a manufacturing organization to produce comparatively small batches of product, as if a continuous-flow line was in operation.
FOCUSED FACTORIES, BUSINESSES, etc.   The establishment of factories or parts of factories and businesses, which are concerned with only one product range, group or market.
GT   Group technology. A production operational layout concerned with the flow of products and materials through the production process. It is based on the assumption that grouping manufacturing resources (such as machines) and making them interdependent has many advantages over functional layouts.
HIERARCHY   The steps in status normally found in a traditional organization.
HYGIENE FACTORS   Factors other than pay, which some social psychologists insist will not motivate people, but which managers often believe can do so, e.g. pensions, health care, time off for family problems, etc.
INTEGRATED DATABASE   The assembly and structure of information which can be used for all core systems in the company.
IT   Information technology. The combined use of computers and communications

technology, to store, manipulate, produce and transmit information both within and between organizations.
JIT   Just-in-time. The process of ensuring that raw materials and work-in-progress arrive at a place and in a time frame which eliminates the need for stock of any kind. It appears to be something of a myth in most British companies.
KANBAN   A Japanese expression related to 'just-in-time' activities.
KBA   Initials of the official German motor industry approval of a product for sale and use.
MANAGEMENT SERVICES   A department which contains computer systems analysts, business studies analysts and programmers, plus O&M personnel.
MARKET SEGMENT   The smallest portion of a total market which will produce a response to a specific targeted marketing activity.
MD   Managing director. A UK alternative to Chief Executive.
MEASURED DAY WORK   A scheme of payments, based on achieving standard output for a predetermined level of pay.
MPS   Master production schedule. This is a key part of carrying out Operational Planning. It is an assessment of order intake compared with capacity available in such detail that delivery promises can be made from it.
MRP and MRP II   Manufacturing Resource Planning. This is a computerized system based on a well-structured database and carries out all the elements of an operational planning activity. MRP II is an advanced version which has an inbuilt feedback mechanism.
MSF   Manufacturing, Science and Finance. A trade union.
MULTI-FACTOR   A scheme usually associated with operative payment where elements other than effort and output are taken into account in assessing earnings.
OEM/OE   Original equipment manufacturer/Original equipment. Motor component suppliers usually service two markets: original equipment (the vehicle assemblers) and replacement.
O/H   Overheads, in accounting terminology. Necessary costs which cannot be directly related to production.
O&M   Organization and methods. An activity somewhat overtaken by systems and business analysis, which sets out to optimize the use of administration resources, by fairly simple analysis and corrective actions.
OPERATIONAL PLANNING   A term used to cover the order processing, master production scheduling, shop scheduling, performance reporting and materials requirements planning activities in a manufacturing company.
OPT   Optimum production technology. This is an approach to planning and using production resources where it is assumed it is best to balance production, not capacity. It suggests that it will always pay to improve bottlenecks' utilization rather than all production facilities.
PBR   Payment by results. A payment system increasingly discredited, where operatives are paid directly for the output they achieve. Most other factors, such as quality, are ignored.
PBT   Profit before tax. The bottom line of the accounts.
PLANNING ELEMENTS   The various components within a planning system

which will help to determine systems design and effectiveness.

PMBD Product master data base. The summation of all information concerned with product manufacturing, constructed in such a way as to facilitate data access.

PRODUCTIVITY The relationship between resources used and output achieved. Of all the measurements that have been proposed:

$$\frac{\text{added value achieved}}{\text{wages paid}}$$

appears the most satisfactory.

PRODUCT MARKET A combination of a product or a product group with a market, which together can be treated as one entity for production, stocking and marketing purposes.

QUALITY ASSURANCE The procedure, methods, technology and attitudes which will ensure the best possible product quality.

QUALITY CIRCLES An activity started in Japan, where operatives are brought together to discuss shop-floor problems (which could be mainly concerned with quality) and propose and perhaps introduce solutions.

ROBOTS Mechanization was the displacement of human muscle by machine. Automation was the displacement of some human thinking capacity by a process controller of some kind. A robot is designed to be a simple, technological replica of how a human being operates.

SCANLON–RUCKER Types of incentive schemes, mainly used in the USA, which use added value, among other concepts.

S, D, & A COSTS Sales, distribution and administration costs, which in a conventional Profit and Loss account are deducted from the Gross Margin to arrive at a net operating profit.

SMMT Society of Motor manufacturers and Traders. A trade organization covering, among other organizations, motor component manufacturers like Ferodo.

SPC Statistical process control. A statistically based activity which needs a clear quality specification. Processes and products are then measured at predetermined times to ensure conformity with the specifications.

SQA Statistical quality assurance. See SPC

STATEMENT OF USER REQUIREMENTS A record of what is happening and what should happen in a major system such as Operational Planning. It should be written by users and then given to systems designers. It should be followed carefully.

STOCK: RAW MATERIALS, WORK-IN-PROGRESS, FINISHED GOODS Together these form the total or gross stocks the company carries. Reducing each element will need a different approach.

TAGUCHI Derived from Dr Genichi Taguchi, who designed a method of quality manufacturing, based on calculating losses caused by factors other than in its intrinsic function.

TQM Total quality management. The philosophy that everyone in a company should be involved with quality.

Apendix 2: List of definitions and explanations of terms and acronyms used

TAYLOR-TYPE INCENTIVES   The incentives devised by F. W. Taylor, based on rating the effort which an operative puts into achieving a task.

UNIONS   In Ferodo we had the following:

- ACTSS   Association of Clerical, Technical and Services personnel. Part of the T&GW Union, covering junior ranks in the clerical work force.
- ASTMS/MSF   At first ASTMS (the Association of Staff, Technical, Managerial and Scientific personnel), later changed to MSF (Manufacturing, Science and Finance). A white collar union which many junior managers and supervisors joined.
- CSEU   Confederation of Shipbuilding and Engineering Unions, of which the old AEU (Association of Engineering Unions) was a member. Mainly concerned with engineering personnel and tradesmen.
- T&GWU   Transport and General Workers' Union. The union normally covering semi- or unskilled operatives.

WIP   Work in progress. All in-process materials and part-finished products which have yet to be completed.

Appendix 3: Training

3.1 Introduction

Those companies which attempt a relevant analysis of their training needs normally find that they have a surfeit of needs and too few resources to attack them all at the same time. What seem like appropriate priorities?

The importance of organizational cohesiveness has been stressed in most chapters of this book. No manufacturing company can ever be run like a well-oiled machine, but there is no excuse for managers not to understand financial results and take action should anything go wrong. The action needs to be unified and coherent. It should not require constant pressure from senior management to activate a suitable response. If the company has to act as a team, then having players fully trained in team tactics is essential. Each player should know that all the other players will respond in a similar way, on a similar occasion. This is the essence of the 'train principle'. Everyone is pulled in the same direction and goes in the same way without complaint.

This, of course, presupposes that management has a clear and well-defined track to go down. Essentially this means an agreed manufacturing strategy is in place, with established priorities, resources allocated and timescales set down. Management should have proved capable of using the manufacturing framework and either singly, but preferably in groups, derived what must be done.

Gaining organizational cohesiveness, then, is of paramount importance. The way Ferodo did this was by submitting everyone, from directors to shop stewards, to a finance course based on past performance, and then repeating it every year while the manufacturing strategy was updated using the manufacturing framework.

This combination of financial awareness and strategic insight produced company cohesiveness. It was reinforced by major quarterly reviews of current results and strategic successes and failures. Perhaps nothing else did more to pull management competence up to levels where being a world-class manufacturer did not seem inappropriate

A3.2 TRAINING PRINCIPLES

Our approach towards training was as follows:

(a) Unless the whole management cadre, plus shop stewards, is trained at the same time and in the same way, using the same information and facilities, it is unlikely that the necessary degree of cohesion will be achieved. Training one or two individuals, in the hope that somehow they will energize everyone around them by their new-found knowledge, is largely a waste of time.
(b) Much can be gained by using teams comprising a mix of shop stewards, supervisors, middle and senior managers, especially when the people concerned originate from a variety of different functions. Perhaps three teams per course might be tried, with some competition in problem-solving between them.
(c) To train in reasonably esoteric topics, it seems essential to use company data and problems as an intrinsic part of the training. Not only will the data and topics take on a real world view, there is also a chance that some of the problems postulated might be edged nearer a solution. Someone should be on hand to record the comments of the teams and where appropriate submit them as amendments or improvements to previously determined strategic and profit plans.
(d) Courses should be held on site during the working week. Any hint of a free-loading occasion should undermine the position of the shop stewards and managers taking part.
(e) Some of the more debatable data strategies and decisions might be discussed more rationally, with an outside professional taking part. Senior management may not always be believed.
(f) The approach outlined is certainly the cheapest way of improving management and perhaps shop steward competence, in the shortest possible time. It is a pity that many training agencies, especially Business Schools, do not show major enthusiasm for helping in on-site training of the type outlined. It seems the wave of the future.
(g) Any number of topics can be covered by the approach, e.g.:

>MRP II
>Financial awareness
>Strategy formulation
>Quality.

Any activity which has company-wide application deserves to be treated in the way outlined.

A3.3 COURSES

(a) Finance
Our finance course covered these subjects:

(i) A review of site performance for the last five years, compared with requisite performance
(ii) The nature of financial and management accounting.
(iii) Financial analysis
(iv) Profit planning—its implementation and a manufacturing strategy.
(v) Monitoring performance and needed response.
(vi) What might be done to improve the chance of achieving or even improving the current profit plan.

Among the individual topics covered were:

(i) The outside world—the impact of environmental pressures (raw material prices, gas, electricity, rates, etc.)
(ii) Price/volume change/capacity utilization
(iii) Trading results
(iv) Cash flow
(v) Capital employed/capital expenditure
(vi) Performance ratios
(vii) Contribution
(viii) Labour and staff costs
(ix) Productivity—definition and ways it can be improved.

Key questions which participating teams had to address included:

(i) 'You have seen the company results both in financial statements and ratios. What do you think are reasonable improvements which can be made on them in the next three years? How is the gap between current performance and the desired objectives to be closed?'
(ii) 'Some parts of the company have done better than others. Some still need to improve more than others. Why? How? (A good question to promote peer group pressure.)
(iii) 'There are some quite significant differences on budgets from, say, three years ago, until now. Even when these are related to potential achievements, some appear out of line. Contact the budget holders and determine why there are discrepancies and what the budget holder intends to do about them.'
(vi) 'What financial statements or evidence do you now need to ensure that the company is on track and doing as well as anticipated?'

(b) Manufacturing strategy
Our course covered these subjects:

(i) A review of the outside world—competition, economic considerations, social changes, changes within the manufacturing framework.

(ii) What kind of a company do we need to be?
(iii) The mission statement. What should it be?
(iv) Leadership—is it good enough?
(v) The manufacturing framework:
 - State of the art, for each element.
 - Appropriate contention and strategies.
(vi) Changes in organization which appear to be necessary.
(vii) Coherent strategy, with due interrelationships.

Amongst the questions teams were asked to consider were:

(i) Has the company an appropriate culture? What should it be? How can it be achieved.?
(ii) Do we understand what it means to be a world-class manufacturer? If so, what parts of current company performance need to be improved?
(iii) Take each element in the manufacturing framework, consider the state of the art and determine where it appears we are deficient.
(iv) Are the current strategies being deployed appropriate to the needs of our product markets? If not, say what else needs to be done. If yes, take two strategies and determine how they should be applied and produce appropriate action plans.
(v) To produce the manufacturing strategy and profit plan, certain assumptions about what is happening in the outside world have been made. Do these assumptions appear valid? If not, what should be substituted?
(vi) The company has limited resources and talent to pursue an appropriate manufacturing strategy. What should be our priorities? How can the whole process be speeded up, without using more resources, especially bank borrowing?
(vii) Each of the elements of the manufacturing framework is important in its own right, but none should be pursued independently of the others. Do we appear to have learned this hard lesson?

Index

activity rates, 179
activity-based costing (ABC), 191
adaptability, 7–8
advanced manufacturing technology, 68–70
aims
   company, 220
   unclear, 113
Allegro, 27
Ambassador Princess, 27
AMT, 69, 75
APV Baker PMC, 69
asbestos, 98, 231–232
Associated Engineering Co., 227
Association of Cost and Industrial
   Management Accountants, 174
ASTMS (MSF) Union, 75, 122, 195
Audi, 25
Automatic Guided Vehicles, 64
automotive components product market
   developments, 29–30

balance sheet and key ratios, 19, 20
bank loans, 85
Bedaux 60/80 schemes, 203
BERAL, 24–25, 27, 39, 215–216
Brigades Union, 213
British Leyland, 27
BTR plc, 25, 83, 216

budgets, 175
   establishing, 186
business information schedules, 183
business reporting reconciliation, 181
Business Week, 24

CAD/CAM, 29, 63, 65
Cahill, John, 83
cash, 9, 85–87, 227–228
   flow, 21
   forecasting, 86–87
   model, 169–170
cohesion, lack of, 114
communication, 133
   and networking in information technology, 150
company analysis, 2–6
competition, 29
competitiveness, 40–41
competitor competence profiles,, 41
Computer-Integrated Manufacturing (CIM),
   63, 64, 75, 231
   benefits of, 68
   difficulty of moving to, 66
   elements of, 65–68
   starting out on, 66–68
conflict, 114

254 Index

and organization in operational planning, 153–155
continuous-flow manufacturing (CFM), 69
contribution and absorption costing, 175–176
control, inadequate, 114
control charts, 53
cooperation, 38
Coopers Lybrand, 158
COPICs, 63
Corporate Mission Statement, 6
COSHH, 2, 77
costing
    contribution and absorption, 175–176
    principles, 186
cost/RPI relationships, 5
costs 8
    administration, reduction in, 87–89
    comparison of, 175
    fixed
        factory, 90–91
        privatization of, 89–90
    internal comparisons, 5–6
    non-manufacturing, 10
    types of, 177
Coventry Climax, 39
CSEU 11, 75, 117–118, 122, 212
culture and developing a manufacturing strategy, 216–220
customers, 8
    relationships with, 48–58
    and sales, 229–230

Deming, William, 55–56
Deming philosphy and quality, 5556
demographic trends, 29, 85
discipline, 126–127, 136–137
    union, 131
discounted cash flow (DCF), 78
DTI, 15
    evaluator report to, 28–29

EC 1992, 38
economic trends, 240
Edwards, Michael, 134
effective organization 119–120
Egan, John, 54
electronic data interchange, 151
Engineering Employers Federation, 144
environmental analysis, 239–240

environmental issues, 29, 231–232
exception reporting, 175

Financial Times 24, 28
finished goods stock (FGS), 212
first-line supervision, 122–132
    in the 1990s 126–131
    objectives and performance measurement, 123–126
    and team working, 127
fixed capacity budget, 175
Flexible Manufacturing Systems (FMS), 29, 64
focus sharpening, 116
Ford Cortina, 27
Ford Motor Company, 6, 49–50, 51, 99, 195, 216
forecasting, 42–47
framework details, 21–22
functional analysis, 88

gap analysis, 18
GEC, 25
General Motors, 79, 80
global economy, 28
globalization, 38
goal conflict, 117–119
Gower, M., 57
Grandage, 41
groups, 120–121

Hanson Trust, 25, 216
Hawthorne Experiment, 197
health and safety, 131
Hegel, 25
Hepworth, 41
Herbert, Alfred, 39
Herzberg, F., 197, 198, 206
hierarchies, 115
high added value, 29, 41
high technology
    and the production process, 62–64
    as answer to manufacturing problems, 78–79
Hitachi, 7
Hoare Govett Investment Research Limited, 215
Hobet Engineering, 69
home market, 39
Honda, 58, 83

Hoover, 69
Hope, Colin, 41

IBM, 69
ICI, 216
IG Metall, 134
incentives, 130
   approach to, 200–202
   end result, 202
   management, 205–206
   needs, 202
   requirements, 201
industry–education relationship, 145–146
information databases, 178
information retrieval, 150–151
information technology, 13
   definition and use, 150–152
   plans for systems, 225–226
   questions concerning, 236
Ingersoll Engineers, 60
innovation, 229
integrated grade/payment system, 212
interfaces, 46
investment, 38
islands of technology, 65, 66, 68, 76

Jaguar, 54
Japanese organizations, relationships with, 57–58
Japanese style of supervision, 131
JCB, 69
job education, 132
Johnsons Wax, 212
Juran, Dr Joseph, 56
Juran philosophy, 56–57
just-in-time, 103–106, 229, 231
   in the office, 89
   stock control, 2
   system, 9, 11

Kant, 25
Kaplan, Robert, 173, 174
key ratios, 19, 20
Komatsu, 39, 212, 213

leadership, 114
limiting factor analysis, 178
line management, 115–116
   involvement in management accounting, 174–178
local area networks (LAN), 150

Lucas, 48–49, 50, 51

management accounting, 13, 152–153, 173–185
   development of, 185–187
   elements of, 178–185
   line management involvement in, 174–178
   material ulilization, 189
   operating statements, 179–185, 190
   potential defects, 187–190
management competence, 232
management incentives/pay schemes, 205–206
management philosophy, inappropriate, 113–114
management resource planning (MRP 2), 11, 13, 42–43, 63–65, 152–153
   activity, 160
   characteristics, 159
   implementation, 170–172
   introduction plan, 170–171
   relationship of local management with, 178
   what users wanted from, 159
management skill, lack of, 113
management staff, 7
management training, senior and middle, 141–142
management/union agreements, 133
Manufacturing Automation Protocol (MAP), 63, 66
manufacturing
   fear of, in the future, 241–242
   strategy, 116–117
   technology, advanced, 68–70
manufacturing strategy table, 14
market share, 39–40
marketing standards, 35
marketing strategic fit activities, 47–48
marketing stategies, 36
Mars, 69, 70
Marx, Karl, 25, 117
Maslow, 197
Massey Ferguson, 69
material flow and technology, 70
materials productivity, 95–97
Matsushita, Konosuke, 7
Mazda, 83
McGregor, 197
Metal Box Co., 103
motivation, 13, 114
motivation systems, 194–214

approach to, 200–202
dilemma in, 194–196
end result, 202
money as only motivator, 197
needs, 202
non-monetary, 197–200
   self-worth, 199–200
plans for, 226
questions concerning, 236–237
requirements, 201
MSF, 75, 122, 195
multi-skilling, 142–146

NALGO, 213
Nautech and Holset Engineering, 69
niche marketing, 37–38
Nissan, 7, 15, 79, 83, 229
Nixon, Sir Edwin, 68
NUPE, 213

objectives
   company, 200
   setting key, 18–19
operational planning, 153–158
   benefits, 156–157
   block diagrams, 162–163
   conflict and organization in, 153–155
   database, 163–167
   design of core elements, 167–170
   desirable improvements in, 157–158
   principles/objectives, 156
   a suitable organization, 155–156
   a uniting function, 155
optimal production technology (OPT), 70, 157, 178
organization structure, confused, 114
organizational change, 38, 116–117
organizational ineptitude, 195
OSI, 66

PA Consultants, 80
pay, 126
   and conditions, harmonization of, 132
   rises, 5
payment systems, 130, 132
   alternatives and problems, 202–205
   future considerations, 212–213
   management, 205–206
   management involvement in changing, 209–210

and motivation, 197
objections to change, 207–209
and relative deprivation, 206–207
solutions, 211–212
way forward, 210–211
PBR, 203
people
   and communications, importance of, 228
   as a resource, 83–85
performance monitoring, 176
personnel, 115
Peters, Tom, 28
philosophy, management, inappropriate, 113–114
pick-and-place units, 64
planning, 130–131
   inadequate, 114
   production requirements, 167–170
   succession, 84
   top-down, 19, 185
plans, establishing, 186
position paper, 11
price rises, 5
problem-solving task committees, 116
product certifications, 36
product design and R&D, 108–109
product market, 12
   questions concerning, 233–244
   plans, 220–221
   requirement, 32–33
product proliferation, 29
product range, 42, 175
production framework, 11–18
production requirements planning, 167–170
profit, 8, 227–228
profit achievement statement, 180
profit and loss account plan sheet, 19, 20
profit planning, 18, 19–23, 241
punishments, 10, 231
   inadequate, 114
purchasing, 101–103
   plans, 224

quality, 230
quality assurance, 48–58
quality awareness, 53–54
quality circles, 2, 54–55, 116
quality standards, 35

rationalization, 38
recording diagrams, 187
recruitment, 126, 229
redundancy, 84–85, 126, 133
relative deprivation and payment schemes, 206–207
reporting procedures, 186–187
research and development, 106–108
　product design and, 108–109
resource usage, 231
resource utillization, questions concerning, 235
resources, 13
　and cost control plans, 223
　data and control, 109–110
response and innovation, 229
Retail Price Index, 5
rewards, 10, 13, 194–214, 231
　dilemma in, 194–196
　inadequate, 114
robots, 64
Rover Group, 39
Rucker/Scanlon added value systems, 202, 204

sales, 8
　and customers, 229–230
scheduling, 130–131
second-tier payment, 212
security, 99–101
senior management, role of, 23–24
shop floor, 8–9
　importance of, 230
simplification, 9
Single European Market, 35, 36
size, 38
social conditions, 240–241
staff (*see also* people), 115–116
standard cost, 175
standard variable cost, 179
standards, 34–36, 175, 176
　action plans on 35
state-of-the-art review, 237
statistical process control, 52–53, 54, 130
stock, 91–95
stock forecasting model, 169–170
stock valuation, 95
strategic fit, 30–31, 45–46, 47–48
　obtaining, 32–35
　and the Single Market, 35

Sturmey Archer, 69
succession planning, 84
systems application architecture (SAA), 150
systems design, 152–153
systems network architecture (SNA), 69
systems specification, 158–162
　outline format, 161–162
systems technology, 13

T&GWU, 11, 75, 117, 118, 122, 196, 212
T&N plc, 215, 227
Taguchi, Dr Genichi, 57
Taguchi process, 57
target achievement, 22–23
Taylor, F.W., 203, 204
team incentives, 211–212
teams, 120–121
　building, 130–131
　monitoring performance of, 128–130
　size, composition, nature, culture, 128
　working, 128–131
technological base of the company, 9, 230–231
technology, 12, 29
　and environment, 239
　and the money problem, 77–78
　plans, 222–223
　questions concerning, 234
teleworking, 151
terms and conditions, harmonization of, 212
testing, 36
Texas Instruments, 212
theft, 99–101
timescales, 240
Tombs, Lord, 1, 22, 41, 77, 85
TOP, 66
top-down planning, 19, 185
Toshiba, 7
total quality control, 2
Total Quality Management, 11, 231
Toyota, 7, 30, 83
trade unions, 2, 134–136
　agreements, 131, 133
　greed, 195–196
　negotiation rights, 132
training, 13, 128, 137–141, 249–252
　and communication, 139–141
　courses, 251–252
　　finance, 251
　　manufacturing strategy, 251–252

    for multi-skilling, 142–146
    principles, 250
    reason for, 138–139
    senior and middle management, 141–142
Training and Enterprise Councils, 144
transitional arrangements, 36

unfair dismissal, 136–137

values, unclear, 113
variances, 179
VW Golf, 27

Walker, David, 173
waste disposal, 98–99

weekly operating statement, 182
Wirkworth Quarries, 127
work organization, 13
    effective, 119–120
    questions concerning, 235–236
    and training plans, 224–225
work simplification, 88
work-in-progress, 91–95
working capital, 9, 91–97
    control sheet, 92–93
world-class manufacturers, 79–80

zero budgeting, 87–88
zone control, 157, 178